T0094244

Hydrodynamic Design and Assessment of Water and Wastewater Treatment Units

Hydrodynamic Design and Assessment of Water and Wastewater Treatment Units

Edmilson Costa Teixeira
William Bonino Rauen

CRC Press
Taylor & Francis Group
Boca Raton London New York

CRC Press is an imprint of the
Taylor & Francis Group, an **informa** business

CRC Press
Taylor & Francis Group
6000 Broken Sound Parkway NW, Suite 300
Boca Raton, FL 33487-2742

© 2020 by Taylor & Francis Group, LLC

CRC Press is an imprint of Taylor & Francis Group, an Informa business

No claim to original U.S. Government works

International Standard Book Number-13: 9781138495890 (Hardback)

Visit the Taylor & Francis Web site at
www.taylorandfrancis.com

and the CRC Press Web site at
www.crcpress.com

Contents

Preface

Water and wastewater treatment has traditionally been an important topic of engineering degree schemes that include *sanitation* in their remit, such as civil, environmental and water engineering. The topic is also of great importance in chemical, mining, oil and gas engineering and, more recently, in the bioprocess and food engineering remits, due to the requirements of related processing plants and/or the need to control emissions of production processes.

Generally speaking, the goal of so-called *treatment* is to modify certain characteristics of the water or effluent, mostly by reducing concentrations of undesirable substances to achieve compliance with legally established standards, and/or adjusting substance levels for subsequent stages of a water or effluent processing plant. For instance, in water treatment for public supply, drinking water quality standards are defined in each country by the Ministry of Health, or an empowered entity acting on its behalf. Also, to prevent or mitigate pollution to natural water bodies, the Ministry for the Environment, or similar, set specific requirements for effluent disposal in rivers, lakes and coastal waters – these are usually much less strict than the drinking water guidelines. Industrial processes each have their own requirements in terms of maximum or minimum acceptable levels of certain substances and elements in their process waters.

Regulations that guide water and wastewater quality control and emission limits are frequently updated to reflect contemporary knowledge on the health and environmental effects of constituents, typically with the overall goals of tackling pollution and adapting to new constraints and discoveries. The general trend tends to be one where regulations become gradually stricter in terms of which substances are allowed in the water or effluent and at what level, the overall aim being to mitigate the risk of both chronic and acute effects that can affect the resilience of human and natural systems in different ways.

Environmental quality is often an important driving force for standard tightening, such as occurs with effluent disposal constraints designed to restrict or prohibit the build-up of certain types of pollution in natural water bodies. Besides episodes which attract much media attention and public commotion, such as a sudden fish die off in a water body affected by an accidental pollutant spillage, there are countless instances of more subtle, slow-building and chronic contamination issues that can have just as serious impacts on ecosystem health and anthropogenic uses – e.g. eutrophication, a relatively common problem in lakes and reservoirs. Socioeconomic consequences can follow if the impacted lake or reservoir were a water supply source for public consumption requiring only conventional treatment, as treatment costs and technology needs could increase to the point of treatment becoming unfeasible. A new, perhaps more distant water source might have to be pursued or, if such an alternative source were not available or also proved unfeasible, then other supply solutions might have to be considered, such as sea water desalination and/or wastewater reuse. Improved watershed management actions could also be implemented, aimed at reducing pollution at the source(s) before unsustainable impact levels are reached and future consequences are worsened. New public policies and more external funding might be required to support such changes if a new, more sustainable, socioenvironmental equilibrium state is to be reached.

In such a broad context, water and wastewater management is constantly balancing demands and supply, while striving to reach compromises between what is desirable and what is possible in view of the current state of socio-environmental affairs, including societal controls and an overall drive to contribute to sustainable development. Likewise, scientific endeavour and professional engineering practice strive to fulfil their core mission of promoting well-being and contributing to the improvement of socioenvironmental conditions. A high level of interdisciplinary integration is required within academia, complemented by a transdisciplinary movement of forming partnerships and working collaboratively, for example with technical staff at treatment plants, operators and managers, to develop and refine tools which have both a sound and up-to-date technical basis and are robust enough to be applicable in day-to-day operations.

Academic degrees that are as wide ranging as most engineering courses have traditionally been designed with such interdisciplinarity in mind – at least as depicted from an analysis of curricula and professional profiling in such areas. It is down to the individual engineer-to-be to inter-relate and integrate knowledge packages delivered in an almost homeopathic way over the years, divided in countless modules of so many different knowledge areas (mostly from the exact sciences but including elements from the life, human and social sciences as well). One such content integration is very much the focus of this book – that between the physics of fluids and the physical–chemical–biological aspects of water and wastewater treatment in an appropriate socioenvironmental context. It is one of the purposes of this book to help bridge that gap in the minds of the engineer-to-be and, generally, in the water and wastewater treatment practice. The book was conceived as a general introductory course to help demonstrate the subject relevance and indicate methods for linking process efficiency with hydrodynamic considerations, thus supporting future design and retrofitting efforts of Water and Wastewater Treatment Units (WWTU). As such, one of the aims here is to highlight the role and applicability of integrative tools that can aid design and retrofit practices, such as conceptual approaches and mathematical equations encompassing aspects of the treatment process kinetics and flow dynamics. It is also expected to contribute towards developing critical thinking on whether, when and how much hydrodynamics really matters in the context of treatment unit design – and where it does, what to do about it. Taking hydrodynamics into account at the design stage of continuous flow systems is arguably always beneficial, even if it is not possible to quantify savings and improvements straight away.

No prior in-depth knowledge of hydrodynamics and associated modelling aspects is required to follow concepts and practical measures associated with the rational design of treatment units, leading to more realistic and potentially improved solutions for new units as well as for retrofitting existing units. Key basic concepts and suitable analytical tools are described, illustrated and worked through using practical examples. Engineering undergraduates and graduates should benefit from the book while undertaking standalone modules on the topic and/or supplementary classes of existing courses on treatment unit processes. The book may also be useful for technical and engineering staff involved in designing and/or retrofitting units for better cost-effectiveness and footprint reduction of the water and wastewater treatment sector.

This book originated from the authors' experience of between 20 and 30 years on the topic, covering lecturing, academic research and industrial applications. The seed text for this book was a module hand-out created and used by the first author, Edmilson Costa Teixeira (ECT), over several years. Many of the ideas and concepts presented and discussed herein were developed and refined as part of research studies and consultancy work carried out by members of the *Hydrodynamics of Reactors Group*, which is currently part of the *Laboratory for Water Resources and Regional Development* (LabGest) of the Federal University of Espirito Santo (UFES), Brazil, formed and led by ECT and from which the second author, William Bonino Rauen (WBR), was a member.

Grateful acknowledgements are due to the Brazilian research funding agencies CNPq (National Council for Research and Development) and CAPES Foundation (Coordination for the Improvement of Higher Education Personnel), which over the years supported a number of research projects mentioned in this book, particularly the authors' doctoral studies and ECT's post-doctoral period at Loughborough University in the UK. We also acknowledge institutional and financial support received from the Brazilian National Health Foundation (FUNASA), Vale S.A., Companhia Siderúrgica de Tubarão (CST) (currently AcelorMittal Tubarão), Espirito Santo State Water Company (CESAN), UFES, University of Bradford (UK) and Hydro-environmental Research Centre (HRC) at Cardiff University (UK).

For the support to publication and dissemination of knowledge in the area of reactor hydrodynamics, particularly with respect to this book, we are very grateful to ABES – Brazilian Association of Sanitary and Environmental

Engineering, AIDIS – Inter-American Association of Sanitary Engineering, ABRH – Brazilian Association of Water Resources, ABCM – Brazilian Association of Engineering and Mechanical Sciences, the Journal of Environmental Engineering / ASCE, the Journal of Hydraulic Research / IAHR and the Chemical Engineering Journal / Elsevier. And for all the attention, patience and professionalism, our deep thanks to Tony Moore and Gabriella Williams of Taylor & Francis Group / UK.

WBR would personally like to thank: ECT for having introduced me to the topic of this book and for his competent and insightful supervision, friendly guidance and continued professional partnership; Eng. Claudio Mattos Machado (currently at ThyssenKrupp) for helping my very first steps in the area of hydrodynamics of reactors; Prof. Binliang Lin (Cardiff University, UK and Tsinghua University, China) and Prof. Roger A. Falconer (Cardiff University, UK) for supervising my Ph.D. in this area (Rauen, 2005) and for all the years of guidance and fruitful collaboration at the HRC/Cardiff University. I dedicate this book to my family and friends, with the deepest appreciation for everything they have done for me, in particular in loving memory of Beatriz who would be very proud to see it coming to fruition.

ECT would personally like to thank: Eng. Arakem M. Oliveira (previously at Geotecnica S.A., currently at GEOHIDRO), who first pointed my way to academia; Prof. Swami M. Villela, EESC-USP, for introducing me to science and indicating me for a Ph.D. at Imperial College/University of London (UK); Prof. Roger A. Falconer, for welcoming me in his research group at the Department of Civil Engineering, University of Bradford (UK) and for proposing my first research project in the area of hydrodynamics of reactors, which led to my doctorate thesis (Teixeira, 1993) – which is where and when it all started; Prof. Koji Shiono, who gave me the opportunity to dive, safely, into the depths of cutting edge science; Mônica M. P. de Almeida, Renato N. Siqueira, Iene C. Figueiredo (all of whom kindly granted permission for reproducing herein material from their respective M.Sc. dissertations and from any other publications derived from them); Bruno P. Vaneli for his general assistance; and, especially, William B. Rauen for bringing this book to fruition; my family, who provides me with support to be the person I am, and in particular my mother Clarice Costa Teixeira and my dear daughter Thalissa Nuttall-Teixeira. Finally, I dedicate this book to my many collaborators and supervised students at the undergraduate (final year and scientific initiation projects) and graduate (masters and doctoral) levels, who provided me with continued learning in the subject area of this book.

About the Authors

Edmilson Costa Teixeira is Professor at the Department of Environmental Engineering of Federal University of Espírito Santo (UFES), Brazil, and has teaching, research and consultancy experience in fluid mechanics, environmental hydrodynamics, sediment transport in hydrographic basins and water resources management and governance.

William Bonino Rauen is Adjunct Professor at the Department of Hydraulics and Sanitation of Federal University of Paraná (UFPR), Brazil, and has teaching and research experience in fluid mechanics, environmental hydraulics, hydrodynamics, solute and sediment transport, as well as water engineering, sustainability and management issues.

A Note on the Perceived Importance of the Topic of This Book in Different Economic Sectors

Some sectors of the economy are more prone to apply scientific and technological innovation than others. For instance, Vaneli (2014) compared socio-economic-environmental conditions experienced by companies in two sectors in Brazil – the water treatment industry and the oil and gas extraction and processing industry – and how their realities influenced the degree of application and further development of hydrodynamic design knowledge.

The oil and gas sector is one of the most competitive, lucrative and potentially polluting industries. Vaneli (2014) assessed whether hydrodynamic design is taken into account in offshore oil extraction plants and, if so, how much. The demand for fluid processing (or 'treatment', by analogy with WWTU processes) involves on-site separation of the oil from salt water, gas and solids that are abstracted from wells, so that the crude oil can be transported elsewhere for further processing and the salty water can be discarded back to the sea as an effluent of the process. The processing complexity, involving multiphase, turbulent, reacting flows is by no means straightforward to deal with and requires advanced techniques and a continuous improvement ethos. Due to the substantial risks to the environment – and largely thanks to past spillages and the public perception of the potential scale of their impacts to ecosystems and local economies – legal standards and reinforcement actions appear to effectively perform their main intended role of preventing pollution. Vaneli (2014) argued that any gain in efficiency, such as that which may be achieved with better hydrodynamic design, can translate into lower unit volumes and/or faster processing and delivery of the oil. In turn, this can lead to substantial cost savings, as space is a premium on offshore platforms and the industry is highly competitive and expensive to run. There is also a lower risk of not meeting regulatory standards for effluent disposal, which translates into less frequent non-compliances and contributes to preserve the company's image. Thus, investment on processing plant improvements is seen precisely as such, with certainty of sound returns. As a result, in this sector in Brazil there is a history of research and development collaboration with academia, including on hydrodynamic design and corresponding technological innovation aimed at improving the separation process efficiency, as well as technical training of company staff.

The water treatment industry is made up of mostly public companies obliged, by law, to meet national potable water standards, as well as collect, treat and dispose of sewage and solid wastes. Their clients are essentially the population of a certain city or region, whom they serve unchallenged by competitors in what effectively works as local or state monopolies. Their primary resource – water – is considered a public good by law (Brasil, 1997) that costs next to nothing. In Brazil, raw water charging is one of the water management instruments prescribed by the National Water Resources Policy (Brasil, 1997). Typical values associated with water abstraction licenses are of the order of 1 cent per cubic metre, an amount which is arguably far from reflecting the true value of water. Thus, extraction, treatment and distribution are the key cost-incurring activities (with pumping, in particular, usually being responsible for the largest share of costs due to the electricity demands). Wastewater reuse is incipient in the country and rainwater harvesting is growing in popularity but is still a timid element of the supply matrix. Water supply is recognised as a public service with an important public health role to play, but the goal of universalising access to potable water has not yet been achieved. There are even larger coverage deficits in the wastewater collection and treatment system (ANA, 2017). The sanitation sector in Brazil struggles to allocate a much-needed fair share of the public funds required to increase sanitation access, year on year. As a result, the investment required to do research and to develop and implement new technology, e.g. associated with hydrodynamic design of WWTU, is far from being a priority – however ethically justified and economically feasible it may be. For instance, Figueiredo (2000) estimated that improving the hydrodynamic performance of a chlorine contact tank, using measures such as those explained in Chapter 3 of this book, could lead to significant cost reductions owing to lower demand for reagents alone. Another important effect would be a lower risk to public health associated with lower disinfection by-product formation.

However, historically the incentives to continuously improve efficiency in the water sector do not seem to have had a strong enough appeal, rather amazingly, as the sector has notoriously been under-funded. An undesired but common consequence of such a situation is a less-than-ideal technical basis for many WWTU in operation, and not only in developing countries (e.g. EPA, 1998). Other aspects take precedence while costs and losses (financial, social and environmental) are ultimately diluted and paid for by the very population who receives below-par services. This situation is comparable to a tragedy of the commons scenario.

There is a significant gap in reinforcement capacity caused, for example, by staff and equipment deficits in regulating agencies, legal loopholes etc. Thus, another important motivator for companies to continuously seek treatment efficiency improvements – a real risk of being penalised for non-conformities – is lost or much reduced. A chronic systemic problem exists that maintains a low overall sewage cleansing efficiency (considering also adequate collection and high-end treatment, and in terms of the organic load) (ANA, 2017), which is the main culprit for the country's notorious low water quality in many water bodies, particularly in urban regions. Among the measures suggested by Vaneli (2014) to improve such a scenario are closer collaborative links between the sanitation sector and research and development institutions, staff training, and – at a macro level – better integration of national policies, particularly those concerning sanitation, water resources and environmental factors. A further suggestion made was to foster healthy competition in the sector, rather than sustaining a monopoly culture, and/or improve regulatory control and reinforcement. Measures such as these can motivate and enable companies to seek continuous improvement of water and wastewater treatment processes, which involves furthering hydrodynamic design considerations, among other aspects.

Introduction

1

1.1 ROLE OF HYDRODYNAMICS IN THE PERFORMANCE IMPROVEMENT OF WWTU

Water and wastewater treatment normally take place in a series of continuous flow units, each designed to perform a step of the purification process – such as the physical removal of suspended material by settling in sedimentation units, or the chemical inactivation of microorganisms by chlorination in contact units – aimed at delivering water or wastewater at a quality that meets the relevant standards.[1] For instance, conventional water treatment usually involves coagulation or flocculation, sedimentation or filtration and disinfection. Each of these processes has its own set of constraints in terms of what is required to achieve the specific treatment goal during the passage of water or effluent through the corresponding unit. Typical factors for consideration include:

1. The characteristics of the water or effluent prior to entering a given treatment unit.
2. The desired characteristics of the water or effluent at the point of leaving a given treatment unit.

---

1 Stricter regulation, environmental constraints and public health issues generally call for continued improvements to water and wastewater treatment systems. While it is technically possible to purify waters or wastewaters to the point of removing all suspended and dissolved elements, this is usually neither economically feasible nor desirable, or necessary from a health or environmental viewpoint, being typically achieved only in laboratories and other applications that require extremely pure water in their processes.

3. The nature of the treatment process required for such modification(s) to occur and what intervention is needed to accelerate or promote it (such as dosage of chemicals, use of aerators etc.).
4. How such intervention takes place (e.g. which reagent should be used, at what dosage rate etc.).
5. Local environmental conditions that may affect treatment, such as temperature, light intensity etc.
6. The rate at which water or effluent passes through the unit and, if the regime is unsteady, what the critical flow conditions to be taken into account are.
7. How water or effluent passes through the unit, in terms of the flow behaviour.

All of the above aspects are usually interconnected, so a change (intentional or not) in one of them may have knock on effects on virtually all the others and on their relative importance. However, usually:

• Factor 1 is more or less set, at least within a certain range, as determined by various factors, for example, the quality of source waters or wastewater produced by a particular industrial process or the outcome of a preceding treatment stage.
• Factor 2 is governed by a fixed treatment goal, such as defined by regulation or as required for a subsequent step of the treatment system.
• Factors 3 and 4 are the two key design aspects of the treatment system and are traditionally considered in view of factors 5 and 6 as well.
• Factor 5 is defined by externalities, although certain control or attenuating measures can be taken if necessary (to mitigate deleterious influences on the treatment process, for example).
• Factor 6 is determined by either the rate of wastewater production or by the rate at which water supply is demanded from a given treatment plant, and can be constrained by how much water can be abstracted from a given source.
• Factor 7 is largely neglected or considered only implicitly in conventional water and wastewater treatment design approaches. It is the main goal of this book to highlight the implications of such a reality, and outline how it does not need to be so, through providing the means to assist with making this factor more widely regarded as a key design aspect of water and wastewater treatment units.

It is common practice to assess the degree to which a given treatment goal is achieved by calculating an 'efficiency' measure of each treatment process that expresses how close (or otherwise) the outflow characteristics are either to the desired characteristics of factor 2, or, relatively to the inflow characteristics of factor 1. Such a measure represents the combined outcome of all of the above listed factors and can be affected by most design and operation actions. During operation of an existing water or wastewater treatment unit, a typical control action to improve such efficiency is to adjust factor 4, as all other aspects are pretty much set for a given operating condition. Design imperfections and/or difficult operating conditions can thus be compensated for, at least in part, and/or temporarily, for example by increasing reagent dosage for a particular process or altering the frequency of aerator usage[2] in a certain unit. However, regular use of such 'medicine' can have deleterious side effects of its own, such as enhancing the production of undesirable by-products and increasing operating costs and energy demands of a given treatment process stage.[3] Not taking factor 7 into account carries precisely this risk and is a common reason why existing water and wastewater treatment units operating in sub-optimal ways need to undergo a retrofit.

It follows that the process efficiency of a given unit depends as much on the physical, chemical or biological reaction of interest as it does on the flow pattern taking place in its interior. This is because the flow pattern governs, for example, the residence/contact time, turbulence levels, collisions and shear to which different fluid portions are subjected to in their passage through the unit. The combined effect of flow features on process efficiency is often overlooked in teaching materials on the design of water and wastewater treatment units. As a result, it is not uncommon to find treatment units operating in a cost-ineffective way, contributing to health problems and environmental impacts, although other factors can also impair process performance.[4]

But the topic itself is certainly not new. Research undertaken in the 1950s in the area of chemical reaction engineering (Danckwerts, 1953; Wehner and Wilhelm, 1956) were scientific precursors of hydrodynamic-

---

2 Where and when used, aerators can significantly affect flow patterns in treatment units, to the point of impairing application of generalised models to estimate hydrodynamic and treatment efficiencies. Aerator usage is not covered in this book.

3 For instance, in water disinfection using chlorine compounds, in addition to inactivating bacteria and viruses, some of the chlorine can also combine and react with dissolved organic and inorganic compounds to form substances which were not originally present in the water (or at least not to the same levels), and which sometime reach levels that have been found to pose harm to human health. Since such an unintended by-product formation was detected, standards have been updated to include such 'new' substances and prescribe maximum acceptable levels in public water distribution systems – a move which has since had knock-on effects on design and control measures of elements of the treatment system, as the water industry was called upon to act.

4 In this book, the terms 'performance' and 'efficiency' are used interchangeably.

kinetic models for continuous flow reactors, whereby the process efficiency of a unit and a hydrodynamic parameter associated with the flow pattern in the unit are explicitly related in analytical form. This was followed, in the 1960s and 1970s, by pioneering work on the hydrodynamics of water and wastewater treatment (e.g. Louie and Fohrman, 1968; Sawyer and King, 1969; Thirumurthi, 1969; Watters, 1972; Kothandaraman et al., 1973; Marske and Boyle, 1973; Thirumurthi, 1974; Hart et al., 1975; Trussel and Chao, 1977; Hart and Gupta, 1978; Silva and Mara, 1979), mostly for chlorine disinfection units and wastewater stabilisation ponds. Subsequent scientific research also focused on other applications including, for example, sedimentation units (Adams and Rodi, 1990; Lyn and Rodi, 1990) and, more recently, a wide range of processes and units have been investigated for the effect of hydrodynamic aspects using mostly Computational Fluid Dynamics (CFD) tools (e.g. Khan et al., 2006; Sartori et al., 2015; Karpinska and Bridgeman, 2016; Meister et al., 2017; Li et al., 2018).

Despite significant improvements leading to so-called 'rational' design approaches (as early as in the 1970s) and modern all-encompassing CFD simulations (mainly since the 2000s), few in-depth applications to other types of water or wastewater treatment units appear to have been translated into design practice. The topic still does not receive too much attention in the remit of typical civil, environmental and water engineering degree schemes and teaching material (apart from a few examples, such as Sykes et al., 2003). Concurrently, significant gaps in knowledge directly applicable to treatment unit design and practice remain to date (e.g. Teixeira et al., 2016; Li et al., 2018).

Thus, perhaps not surprisingly, a large number of existing water and wastewater treatment units operate in a non-optimal way, i.e. generally their operation does not meet, concurrently, the following conditions:

1. Maximisation of treatment process efficiency.
2. Minimisation of operation costs.
3. Minimisation of undesirable by-product generation.

Condition number 3 (minimisation of undesirable by-product generation) has conferred, since the pioneering studies of the 1960s, a new meaning to the optimisation of water and wastewater treatment processes, either due to environmental concerns or public health. All three causes are related to flow features in WWTUs.

Among the contributing factors for low performance of units (from the viewpoint of the optimisation of treatment processes) are inadequate design and operation procedures. A few rule-of-thumb design guidelines with underlying flow hydrodynamics reasoning can be found for some types of WWTUs, as will be shown later in this book.[5] However, usually little or no advice is available in such guidelines when things go wrong, when adaptation is required due to exceptional project requirements or when there are unforeseen circumstances to deal with.

Taking the process of water disinfection for public supply as an example, we have:

1.1.1 Inadequate Design Procedures

- False hypothesis that flow behaviour in chlorination units is plug flow, while flow patterns observed in practice are considerably more complex and characterised by significant short-circuiting and dead and/or recirculating zone occurrence.
- Insufficient knowledge about the effect of treatment unit setup on the disinfection efficiency. The flow pattern in a treatment unit can be highly influenced by its setup (geometry, type of inlet and outlet devices, flow deflectors, baffles, etc. – in terms of the presence or absence of such elements, as well as their quantity and arrangement).
- Absence of suitable injection devices that promote a uniform distribution of the disinfectant agent at the inlet section of contact units.

1.1.2 Inadequate Operation

Due to inadequate design procedures, as outlined above, water treatment plant operators tend to use higher disinfectant dosages than theoretically necessary to achieve a given disinfection efficiency level. Thus, the treatment unit operating costs are increased while the formation of by-products is favoured – by-products that are potentially harmful to human and environment health, as occurs with the formation of carcinogenic compounds when water disinfection uses chlorine compounds.

---

5 Concepts of the area of fluid dynamics can also be applied in other areas where physical, chemical and/or biological reactions take place in flowing fluids inside closed vessels. For example, the internal configuration of combustion chambers in an engine can be optimised to improve performance, as perceived in terms of maximising energy conversion (e.g. from chemical to mechanical) while minimising undesirable by-product formation (such as avoidable polluting gases associated with combustion inefficiencies).

1.2 INTRODUCING HYDRO-KINETIC MODELS FOR HYDRODYNAMIC WWTU DESIGN

Several examples, in a range of environmental sanitation areas, can be used to demonstrate the importance of the hydrodynamics of reactors for the optimisation of water and wastewater treatment processes.[6] Examples of equations that involve a fluid dynamics parameter to help estimate the efficiency of water and wastewater treatment processes – so called hydro-kinetic models – are discussed herein. All such equations can be regarded as simplified descriptions of the complex, transient, three-dimensional physical, chemical and/or biological processes taking place in continuous flow reactors, sometimes with interactions between them. Mathematical models that more closely represent certain treatment processes can be directly tackled through advanced CFD modelling approaches (e.g. Zhang et al., 2000; Angeloudis et al., 2014a, 2014b; Meister et al., 2017), thanks to contemporary computing power and numerical modelling techniques being well beyond the corresponding capability in a not too distant past. More complex and usually multiphase flow conditions remain a modelling challenge (e.g. Karpinska and Bridgeman, 2016), due mainly to the incomplete mathematical description of such phenomena. Thus, and as advanced CFD modelling is still beyond the bulk of the treatment unit design and operation practice, relatively simple equations and 'pragmatic' approaches able to provide significant improvements to process efficiency should have a role to play for some time to come (e.g. Shilton et al., 2008; Teixeira et al., 2016).

A relatively simple mathematical approach in this sense is to take into account the effects of longitudinal dispersion and mixing in the unit by means of a one-dimensional (longitudinally variable) modelling approach, i.e. one that assumes cross-section uniformity of variables such as the diffusion coefficient and sub-stance concentration. Further simplifications are often also used, such as assuming steady state flow conditions (and perhaps simulating critical scenarios) and parameterising any effects that are not explicitly represented in the functional

---

6   This book does not cover treatment unit design approaches that ignore the fact that the flow field in continuous flow units is non-ideal, sometimes departs greatly from an ideal reference condition, and that this can have a significant impact on process efficiency. This is the case of so-called batch kinetics models (e.g. Chick-Watson's disinfection model), for which the literature is abundant for the interested reader to consult.

relationship thus created. This type of approach led to the proposition of design standards for treatment units where, and when, sufficient knowledge of the effects of hydrodynamics on treatment unit efficiency was available (e.g. EPA, 1983, 1999). As a result, typically one hydrodynamic parameter is explicitly included as part of process efficiency-related equations, as exemplified below.

Firstly, we calculate the disinfection efficiency (*DE*) as:

$$DE = 1 - \frac{C_{out}}{C_{in}} \tag{1.1}$$

where
   DE = disinfection efficiency, which can theoretically lie in the range of 0 (no disinfection) to 1 (full disinfection);
   C_{in} and C_{out} = pathogen concentrations at the inlet and outlet sections of the WWTU respectively.

The ratio C_{out}/C_{in} can be estimated in a variety of ways. Equation 1.2 was developed by Wehner and Wilhelm (1956) and has been used since the 1960s to design and estimate the efficiency of wastewater stabilisation ponds (Keller and Pires, 1998). For example, Sarikaya and Saatçi (1987) used it to describe micro-organism disinfection efficiency; Crites et al. (2005) reports its usage to estimate the removal of biological oxygen demand in facultative ponds; Teixeira and Sant'Ana (1999) applied it to simulate biological filters; and Teixeira et al. (2016) tested its ability to describe the disinfection efficiency of contact units for water disinfection. The latter study found Equation 1.2 gave lower estimation errors than the US industry standard for disinfection unit design (i.e. the *C* −*t* rule, discussed below) under a range of hydrodynamic and kinetic conditions. However, Teixeira et al. (2016) also pointed out that care should be taken when applying such an equation to high mixing conditions, as it was not originally developed for such, and to not use it as the sole design tool of water disinfection units since it was less conservative than the *C*–*t* rule and could lead to health risks.

$$\frac{C_{out}}{C_{in}} = \frac{4\, a\, e^{\frac{1}{2d}}}{(1+a)^2\, e^{\frac{a}{2d}} - (1-a)^2\, e^{-\frac{a}{2d}}} \tag{1.2}$$

where
   $a = (1 + 4kTd)^{1/2}$;
   d = dimensionless coefficient that characterises the level of mixing in the reactor and comprises a 'hydrodynamic link' in the equation;
   e = 2.71828 is the Naperian number;

T = theoretical residence (or retention) time (T= V/Q), a so-called *hydraulic* parameter[7];

V = working volume of the reactor;

Q = flow rate; and

k = first order kinetic reaction rate constant.[8]

Equation 1.2 makes explicit the dependency of the treatment process's efficiency on hydraulics (T), reaction kinetics (k) and reactor hydrodynamics (d).

The so-called C–t rule (EPA, 1999) is an adaptation of a well-known batch disinfection model, namely the Chick-Watson (C-W) model (e.g. Gyurek and Finch, 1998) for continuous flow contact units. The letter C stands for the concentration of disinfectant (such as chlorine), while t stands for a reference contact time between such disinfectant and a microorganism (such as *E. coli*). In the C-W model, such a variable simply reflects the experimentation time in a batch reactor. If directly applied to a continuous flow unit, t is taken as the theoretical residence time in the unit (T) (T = V/Q, where V is the unit volume and Q is the flow rate) but this approach normally leads to large errors in (over) estimating the disinfection efficiency, particularly as the flow pattern deviates from plug flow (e.g. Teixeira et al., 2016). Thus, in the C–t rule, t is taken as the hydraulic-hydrodynamic parameter t_{10}, which represents the time for 10% of a substance's mass entering the unit at any given moment to leave the unit (pass through its outlet section).[9] Equation 1.3 is used to determine the concentration ratio:

$$\frac{C_{out}}{C_{in}} = e^{-k'C^n t} \tag{1.3}$$

where

k' = is a microorganism constant; and

n = is a dilution-related coefficient.

Equation 1.4 was developed by Trussel and Chao (1977) and includes the residence time distribution (RTD) curve $E(\theta)$, which describes the temporal

---

7 See Section 2.2 for comments on the usage of the terms 'hydrodynamic' and 'hydraulic' in this context.

8 A number of models exist to describe reaction rates as impacted by few or several control variables. Being outside the scope of this book and for the sake of brevity, it can be assumed herein that an appropriate representation of the reaction kinetics is made. The treatment unit can then be thought of as a sequence of batch reactors in which the reaction kinetics occurs according to k.

9 As such, t_{10} is one of the most important hydrodynamic efficiency indicators, as explained later.

variation of concentration at the outlet section of a substance injected at the inlet section.[10] It is often referred to as the segregated flow model for its ability to take into account the variation of key process variables along the flow through a unit. It was the basis for the later development of the Integrated Disinfection Design Framework (IDDF) (Bellamy et al., 1998).

$$\frac{C_{out}}{C_{in}} = \frac{1}{T} \left(\int_{0}^{t_d/T} E_\theta \, d\theta + \int_{t_d/T}^{\infty} \left(\frac{t_d}{C_0 \, T \, \theta} \right)^3 E_\theta \, d\theta \right) \tag{1.4}$$

θ = contact time normalised by the theoretical residence time ($\theta = t/T$);
   t_d = required time for a disinfectant to diffuse through a microorganism membrane and effectively start disinfection;
   C_0 = initial concentration of residual combined chlorine; and
   E_θ = normalised residence time distribution curve for the reactor.
   The relationship between the theoretical E_θ curve with the mixing coefficient d is given by Equation 1.5, but this can also be measured in field experiments or through laboratory studies, as detailed later in this book.

$$E_\theta = \frac{e^{\frac{-(1-\theta)^2}{4d\theta}}}{\sqrt{4 \pi d \theta}} \tag{1.5}$$

It is noteworthy that relatively resistant organisms and/or small units and/or high flow rates all tend to make t_d approach T. As a result, the ratio C_{out}/C_{in} grows and, thus, DE decreases. This is because the importance of the first integral on the right-hand side of Equation 1.4 (which contributes nothing to the disinfection efficiency as the disinfectant agent only starts acting once inside the organism) grows in relation to the second integral (which effectively represents disinfection taking place). The opposite effect is predicted by Equation 1.4 in cases of disinfection of less resistant organisms (for a particular disinfectant) and/or relatively large units and/or low flow rates.

In any case, the unit hydrodynamics has a key role to play in the overall DE obtained, as exemplified above. A direct link with the efficiency of water and wastewater treatment processes exists and is explicit in equations such as 1.2, 1.3 (with $t = t_{10}$) and 1.4. Even though this link can be apprehended and accepted, it is necessary to know, then, how a more detailed analysis of

---

10 The RTD is a key tool to assess the hydrodynamic efficiency of flow reactors, as also explained later.

reactor fluid dynamics can be useful in the areas of design and operation of water and wastewater treatment units. Possible approaches include:

a) Diagnostic studies of a unit's hydrodynamic performance and its influence on the process treatment performance.
b) Proposition of corrective measures (such as changing a unit's setup, flow rate, etc.), which can contribute to improve the treatment unit performance.

Both approaches can be applied to improve performance of either existing or planned units, with the purpose of optimising the water or wastewater treatment process,[11] as shown in this book.

A large number of equations is available in the technical and scientific literatures for use in the design of continuous flow WWTU, and/or to estimate their process efficiency, which do not explicitly include a hydrodynamics-related parameter such as t_{10} or d. However, many such equations include geometrical parameters such as the flow depth, width or length-to-width ratio, sometimes in addition to T. When combined with design criteria and recommendations associated with, for example, specific inlet or outlet characteristics, depth values, usage of baffles to create serpentine-like flow paths etc., then these geometrical parameters may, at least in part, represent the hydrodynamic efficiency as surrogates. Evidence of this observation can be found in publications such as Von Sperling (1996a, 1996b) and Howe et al. (2012), which provide compilations of relationships that link d, in particular, to geometrical characteristics of WWTU and, sometimes, also to specific types of WWTU assuming typical constructive features. By combining such geometrical-hydrodynamic equations with appropriate geometrical-process efficiency equations, and after some algebra, one can easily generate a vast number of alternative hydro-kinetic models to estimate the efficiency of processes of interest considering the effects of hydrodynamics.

---

11 Assessments such as proposed herein could involve a hydrodynamicist and a treatment process specialist, with perhaps a health professional, an economist, a limnologist, a manager of water or effluent treatment plant, a regulator and a civil engineer. The main aim of such a team could be to identify integrated and cost-effective solutions for real problems facing water and wastewater treatment systems. Perhaps not surprisingly, they might find that scientific advances and capabilities in some disciplines are far ahead current treatment practice, as there have not been enough attention and resources dedicated in research grants and projects, over the years, for joint problem definition and solution implementation with technical staff and operators in the water sector. If any such 'ideal' solution were to be found by the group, it would probably not be the cheapest one that a short-termist economist could think of, nor the most optimal one from the perspective of any one of the other professionals, but it would likely strike a balance between the various needs and possibilities of all areas, and perhaps prove to be the most cost-effective solution in the medium to long term.

Fundamental Concepts and Techniques

2

Presented herein are some key definitions which will enable a better understanding of topics discussed in this book. Section 2.1 contains a review of concepts related to solute transport and mixing processes in water bodies. Section 2.2 presents key definitions in the area of hydrodynamics of reactors, with focus on flow pattern classification and features in water and wastewater treatment units (WWTU). Section 2.3 explains how such flow patterns can be assessed and analysed. Other, more specific definitions are presented in subsequent chapters as needed.

2.1 SOLUTE TRANSPORT, DISPERSION AND MIXING PROCESSES – AN OVERVIEW

We start this section by drawing upon fluid flow classification aspects which are normally covered in fluid mechanics textbooks (e.g. Pritchard, 2011; White, 2011). For instance, flows can be classed as steady or transient with respect to the behaviour in time; uniform or non-uniform with respect to spatial distribution; laminar, transitional or turbulent with respect to the turbulence level; and so forth. We also recall the concepts of streamlines, vector fields and isocontour plots, which form an analytical tool basis to describe and study flow fields and identify the occurrence of key features, such as flow separation and reattachment, high and low speed zones, accelerations, pressure gradients, etc. It is also worth remembering that flows are usually strictly three-dimensional (3D) – i.e. the velocity vector at any given point has non-zero components in all spatial directions – but under certain circumstances it is possible and sufficient to disregard one or two components and describe the flow as two-dimensional (2D) or one-dimensional (1D) in nature respectively.

With such aspects and concepts in mind, we offer definitions for technical terms adopted herein, mostly in line with Fischer et al. (1979) with regard to classifying transport phenomena in water bodies.

Convection: fluid movement caused by temperature gradients; a type of density current, which involve the movement of denser fluid parcels (pulled down by gravity) concomitantly with the displacement and upwards movement of less dense parcels situated below; density change in single phase fluids is usually caused by temperature discrepancies in layers or regions of flow; may occur in the absence of an externally driven flow field and potentially cause internal streaming and recirculation.

Advection: fluid movement caused by the mean current, rather than by turbulent eddies, as considered during a finite time window.

Shear advection: streamwise fluid movement caused by currents under a normal velocity gradient, usually in the vicinity of a solid boundary or a recirculating flow structure (i.e. non-uniform cross-sectional flow); as a result, fluid parcels situated in a higher speed flow region area advected faster than neighbouring fluid parcels situated in a lower speed flow region, causing separation of such flow parcels as advection takes place. Due to shearing action, micro-vortexes are formed that cause transversal spreading of momentum and solutes, thus contributing to turbulence generation and substance mixing.

Molecular diffusion: fluid spread caused by random molecular movement, usually 3D and isotropic; intensity represented by the molecular diffusion coefficient (D); usually not very perceptible in flowing fluids as occurs at a lower rate than other processes described herein (molecules take a long time to travel macroscopic distances!).

Turbulent diffusion: fluid spread caused by random turbulent (eddy) movement, usually 3D and anisotropic; intensity represented by turbulent (eddy) diffusion coefficient (ε), which is typically orders of magnitude higher than D; usually occurs together with advection, i.e. in flowing fluids, except under controlled laboratory conditions with an oscillating grid placed in an otherwise still fluid (e.g. Honey et al., 2014).

Dispersion: fluid spread caused by shear and diffusion; thus, dispersion has a random component (diffusion) and a directed component (advection).

Mixing: generic term used to represent any type of fluid spread, regardless of the cause; for instance, flow recirculation, convection and dispersion can all combine to cause mixing.

A widely used and simplified way to represent an engineering flow (e.g. in a river or pipe) taking place between an upstream section and a downstream section, without describing in detail what happens in between them, is to regard it as 1D, non-uniform, turbulent and steady for a given scenario of interest. In the absence of temperature discrepancies, convection can be disregarded as a spreading process, as can molecular diffusion due to it being relatively unimportant compared to turbulent diffusion. This leaves shear advection and turbulent diffusion as the key causes of fluid spread along the flow region under consideration, which is a typical case of dispersion, for example in free surface flows. The amount of spread experienced by an inflowing fluid parcel as it moves from one control section to the other can then be represented in terms of the longitudinal dispersion coefficient (D_L).

Many authors have developed empirical and semi-empirical relationships for D_L involving influencing parameters such as the mean flow speed (U), width (B) and depth (H), among others. For example:

$$D_L = 5.93 H u_*$$ (2.1)

(Elder, 1959)

$$D_L = \frac{0.011 U^2 B^2}{H u_*}$$ (2.2)

(Fischer et al., 1979)

where u_* is the friction velocity. There exist several other D_L equations that usually give different results between them, with the discrepancies sometimes spanning orders of magnitude (especially if a given equation is applied outside its scope of development). For instance, Elder's equation consistently tends to underestimate D_L by one or two orders of magnitude relative to other D_L equations and to actual amounts of dispersion in rivers, but it performs better for artificial wide channel flows (Fischer et al., 1979; Pereira and Teixeira, 2002).

The overall amount of spread caused by dispersion along a flow path of length L can be estimated using the dimensionless coefficient (d_L), a ratio between longitudinal dispersion and advection transport rates in a flow:

$$d_L = \frac{D_L}{UL}$$ (2.3)

This, as pointed out by Levenspiel (1999), is similar to – but not exactly the same as – the reciprocal of the Peclet number (for heat transfer) or the Bodenstein number (for mass transfer), both of which involve the molecular diffusion coefficient (D) rather than D_L.

Where other fluid spreading processes (such as convection, recirculation etc.) take place between the two control sections, it is, by definition, appropriate to describe the total amount of spread experienced by an inflowing fluid parcel in terms of the global mixing coefficient (d). This is the very coefficient that appears in the hydro-kinetic Wehner and Wilhelm model, Equation (1.2), as seen in Chapter 1. Under any circumstances it is, by definition, correct to say that $d \geq d_L$ with the difference between them representing all other causes of mixing, other than dispersion. In any case, this type of approach lumps together into a single parameter the effect of all flow processes on fluid spread between two control sections – so called parameterisation of mixing effects. In the following section we will see how this can be useful in describing flow patterns in continuous flow reactors, such as water and wastewater treatment units.

2.2 TYPES AND FEATURES OF FLOW PATTERNS IN CONTINUOUS FLOW REACTORS

Continuous flow reactors are treated as closed vessels herein, as defined by Levenspiel (1999). This means that they comprise of well-defined, wetted volumes surrounded by lateral solid boundaries and a solid bed, often with a free surface and sometimes working as a pressurised conduit, such as occurs in pipe reactors. These vessels also have well defined inlet and outlet flow structures/devices through which there is no return of passing fluid parcels. Once inside the vessel, fluid parcels follow flow paths from inlet to outlet, and the framework formed by the combination of such flow paths is known as a flow pattern. Different types of flow patterns are considered in the study of hydrodynamics of reactors, particularly in reactor-wide 'black box' analysis approaches focused on the overall effects of fluid passage from inlet to outlet, i.e. without analysing in detail what happens in between such sections.

Real flow patterns: flow patterns that combine features of the idealised types of flow pattern (*plug flow* and *complete mixing*);

Idealised flow patterns: flow patterns which are not observed in practice (eg., plug flow and complete mixing), but which each represent an ideal design condition in the context of a given treatment process.

Complete mixing: idealised flow pattern in which throughout the length of the reactor the invariability of fluid properties (mass density, constituents concentration, etc.) is verified at any time.[1] It is the recommended flow pattern, for instance, when treating wastewaters subjected to highly variable and/or toxic loads, due to promoting rapid and efficient equalisation in the unit (von Sperling, 1996b). Types of WWTU typically associated with this flow pattern include, for example, equalisation units in general and anaerobic digesters used in nutrient removal (e.g. Raboni et al., 2014; Capodaglio et al., 2016).

Plug flow: idealised flow pattern in which all fluid parcels entering the unit at a given time have equal velocity and follow parallel and straight paths from the inlet to the outlet sections of the unit. Thus, parcels take the same time to go through the unit, or in other words, all parcels have equal residence times (T_{pf}) in the unit. This flow pattern is particularly relevant in treatment processes involving first order kinetics (von Sperling, 1996a), being a prevalent design assumption for WWTU such as sedimentation and contact units.

Figure 2.1 illustrates the idealised flow patterns of plug flow and complete mixing, and provides an example of a real flow pattern. Under complete mixing it is assumed that a fluid parcel entering the reactor via the inlet section instantly mixes with the fluid already in the reactor. At the outlet section, the concentration of such fluid is always the same as inside the reactor and, as time progresses, such concentration gradually reduces until all trace of that fluid parcel is eliminated from the reactor.[2] Such concentration reduction as a function of time occurs due to dilution of a conservative substance by new fluid entering the reactor continuously. Under plug flow, on the other hand, transport occurs only due to advection and, it is assumed, uniform inviscid laminar flow in the reactor, with the absence of shear, diffusion and any mixing-causing effects.

Plug flow with dispersion: a best case scenario real flow pattern in which the inevitable occurrence of non-uniform, turbulent flow is taken into consideration to predict the minimum level of dispersion for the reactor (i.e. effects of shear advection and turbulent diffusion) (e.g. Falconer and Tebbutt, 1986). As such, this flow pattern represents a theoretically possible condition, which could be achieved if other causes of mixing (such as recirculation), apart from dispersion, could be eliminated from the flow pattern. In hydraulic optimisation studies of water and wastewater treatment units where plug flow is the ideal condition, this flow pattern represents a more realistic target for improvements than plug flow, as shown in Section 2.3.

---

1 An underlying consideration made here is that such instant and continuous homogenisation occurs as result of flow features in continuous flow reactors, and not due to aerator usage or mechanical agitation as in a batch reactor.

2 Strictly speaking the concentration reduction becomes asymptotical with respect to time.

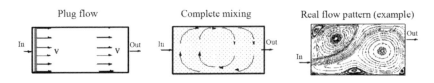

FIGURE 2.1 Schematic representation of the idealised flow patterns: plug flow and complete mixing (adapted from Almeida, 1997); and an example of a real flow pattern

Theoretical residence time (T): the theoretical residence time for a reactor is the time that all fluid parcels would remain in the reactor if the flow pattern were plug flow (Teixeira, 1995a), as illustrated in Figure 2.2. It is calculated as the ratio between working volume and flow rate, i.e. $T = V/Q$. This is frequently referred to as the hydraulic residence time (HRT) and is commonly used for WWTU design purposes.

Residence Time Distribution (RTD) – E(θ) curve: relative distribution of residence times of fluid parcels that entered a unit in a small time interval, for a given hydraulic condition. Figure 2.3 illustrates three RTDs, which are related to each of the idealised flow patterns and a hypothetical real flow pattern. In the figure, the horizontal axis is the normalised time θ, such that $\theta = t/T$, where $\theta = 1$ indicates that $t = T$; and the vertical axis is the normalised outlet concentration $E(\theta)$ of a conservative substance injected instantaneously at the inlet section at $t = 0$, where such normalisation is conducted as:

$$E(\theta) = \frac{C(\theta)}{C_0 REC} \qquad (2.4)$$

$C(\theta)$ is the outlet concentration as a function of θ, *REC* is explained in the following paragraph, and C_0 is the bulk average substance concentration inside the unit, i.e. $C_0 = m/V$, where m is the instantaneously injected substance mass and V is the wetted volume of the treatment unit. Outlet monitoring of substance concentration following a controlled inlet injection is the most commonly used method to acquire RTDs but other approaches can also be used, as discussed in Section 2.3. Fischer et al. (1979) and French (1985) show comparable applications in natural water bodies, while Levenspiel (1999) covers in-depth material on RTD acquisition and usage to infer on mixing processes in chemical reactors, such as in many WWTU.

In WWTU tracer tests it is a relatively rare occurrence to detect exactly 100% of the substance mass injected at the inlet section, even with conservative tracers. Typical acceptable 'recovery' (*REC*) values range between 85% and 115% (Stamou and Adams, 1988), also due to equipment and sampling limitations. The division by *REC* in Equation 2.4 consists of a correction step required at the stage of data processing so that any *E(θ)* curve obeys the principle of mass conservation – which is a key pre-requisite for subsequent comparative analyses. Calculating *REC* involves computing how much tracer mass was detected through monitoring, such as:

$$REC = 1/m \int_{0}^{\infty} Q(t)C(t)dt \tag{2.5}$$

The plug flow RTD representation in Figure 2.3 consists of a narrow and steep curve, such as given by the Dirac delta function (e.g. Fischer et al., 1979, p. 39), which leads to:

$$E(\theta) = \begin{cases} +\infty, & \theta = 1.0 \\ 0, & \theta \neq 1.0 \end{cases} \text{for the plug flow pattern} \tag{2.6}$$

On the other hand, the complete mixing curve starts at *E(θ)* = 1.0 for *θ*= 0 and decays exponentially with *θ*, such that:

$$E(\theta) = e^{-\theta} \text{for the complete mixing flow pattern} \tag{2.7}$$

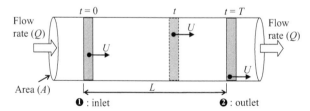

mean cross-sectional flow velocity: $U = Q/A$
hydraulic residence time: $T = V/Q = AL/Q = AL/UA = L/U$

FIGURE 2.2 Schematic representation of idealised plug flow in a pipe and implications for mean velocity and hydraulic residence time calculations

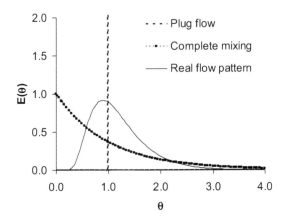

FIGURE 2.3   Residence Time Distributions of the two idealised flow patterns (plug flow and complete mixing) and a hypothetical real flow pattern

The RTD curve of a real flow pattern always falls somewhere in between the curves of the two idealised conditions, due to the fact that the real flow pattern contains features of both idealised cases to a certain degree; the closer a curve associated with a real flow pattern is to that of one of the idealised flow patterns, the higher the degree of resemblance of flow features is between them, from the viewpoint of residence times of fluid particles.

It is also noteworthy that, for all $E(\theta)$ curves:

$$\int\limits_{0}^{+\infty} E(\theta)d\theta = 1.0 \tag{2.8}$$

which reflects the principle of mass conservation of the injected substance, from inlet injection to the outlet passage, represented by the Residence Time Distribution curves.

Cumulative RTD for a reactor – F(θ) curve: represents the accumulated substance mass that has passed through a given monitoring section since injection, up to the normalised time θ, as calculated by Equation 2.9. Figure 2.4 illustrates $F(\theta)$ curves associated with plug flow, complete mixing and a hypothetical real flow pattern.

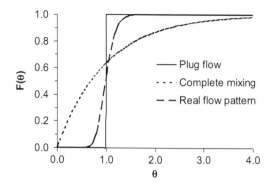

FIGURE 2.4 Cumulative passage curves for a hypothetical real flow pattern and the idealised plug flow and complete mixing flow patterns

$$F(\theta) = \int_0^\theta E(\theta)d\theta \qquad (2.9)$$

where, for plug flow, $F(\theta) = 0$ as $\theta < 1$ and $F(\theta) = 1$ as $\theta \geq 1$; and for complete mixing, $F(\theta) \to 1$ as $\theta \to \infty$, reflecting the asymptotic behaviour of the tail of the corresponding $E(\theta)$ function.

In practical terms, in many WWTU tracer studies, for nearly conservative substances and due to detection limits, $F(\theta) \approx 1$ is typically reached at θ values above unity and up to approximately four. In flow patterns tending to plug flow, and for all practical purposes, full substance passage can often be seen to occur for θ up to approximately two.

Key flow features occurring in water and wastewater treatment units are described below.

Reversed flow: flowing parcels directed contrary to the main streamwise direction in a given cross section.

Recirculation zone: loop-like trajectories followed by fluid parcels around a slowly rotating centre; fluid parcels that enter recirculation zones in a treatment unit tend to exhibit longer residence times than T, particularly if trapped near the centre of such a zone where the escape probability is lower than towards the outer rotating region.

Dead zone: flow region in which fluid parcels remain for much longer periods, on average, compared to T. Such regions are characterised by near-zero exchange rates (e.g. of heat and mass) with neighbouring flow regions, as the fluid is practically stagnant (Teixeira, 1995a).

Short-circuit: pathway of fluid parcels which lead to lower residence times than T; it is noteworthy that relatively short distances between the inlet and outlet sections (such as straight lines) are not necessarily associated with the occurrence of a short-circuit pathway, as defined herein, for there must be a flow streamline associated with a short-circuit route (Teixeira, 1995a).

The RTD curves illustrated in Figures 2.5 and 2.6 refer to flow patterns with low and high intensities of short-circuit respectively. It can be noted that the Residence Time Distribution curves of Figure 2.5 are narrower than the ones shown in Figure 2.6. While the distributions shown in Figure 2.5 represent a flow pattern tending to plug flow, the distributions shown in Figure 2.6 approach the curve associated with the complete mixing flow pattern. Later it will be shown how high short-circuiting intensity is directly linked with the presence of substantial reactor volumes occupied by dead zones and recirculating flows.

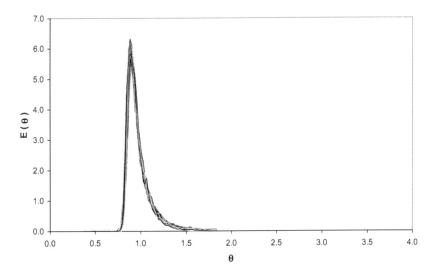

FIGURE 2.5   Example of Residence Time Distribution curves (flow pattern with low short-circuit intensity) (adapted from Siqueira, 1998; Teixeira and Siqueira, 2008)[3]

3 Republished with permission of American Society of Civil Engineers, from *Journal of Environmental Engineering*, Performance Assessment of Hydraulic Efficiency Indexes, Teixeira, Siqueira, 134 (10), 851–859, 2008; permission conveyed through Copyright Clearance Center, Inc.

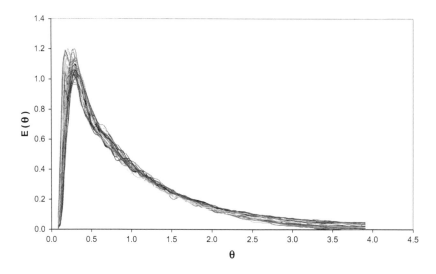

FIGURE 2.6 Example of Residence Time Distribution curves (flow pattern with high short-circuit intensity) (adapted from Siqueira, 1998; Teixeira and Siqueira, 2008)[4]

Hydraulic efficiency or hydrodynamic efficiency: a measure of the level of proximity between the ideal flow characteristics for a given reactor and the flow characteristics observed in practice (Teixeira, 1995a).

In WWTU literature, particularly from the unit processes perspective, the term 'hydraulic' is generally associated with flow-related aspects that somehow influence the treatment efficiency, with the HRT being the iconic 'hydraulic' parameter. As the RTD concept was introduced in WWTU design and assessment studies, and starting with physical hydraulic models (e.g. Falconer and Tebbutt, 1986), the term 'hydraulic efficiency' (Teixeira, 1993) has been coined and widely used to flag and represent the influence of such flow-related aspects on WWTU performance, much more than 'hydrodynamic efficiency' (including by these authors). Both terms convey the same information as reported in WWTU literature. The term 'hydraulics' is perhaps more easily recognised and generally understood by a non-hydrodynamicist audience, such as traditional process specialists and technical

---

4 Republished with permission of American Society of Civil Engineers, from *Journal of Environmental Engineering*, Performance Assessment of Hydraulic Efficiency Indexes, Teixeira, Siqueira, 134 (10), 851–859, 2008; permission conveyed through Copyright Clearance Center, Inc.

staff in the water industry. However, by realising that even a simple RTD is typically the combined result of several fluid flow (hydrodynamic) processes that govern substance spread and transport in a WWTU (such as turbulent diffusion, shear advection and pressure gradients associated with flow separation and reattachment), it seems factually more correct to use the term 'hydrodynamic efficiency' *in lieu* of 'hydraulic efficiency'. The former, arguably more precisely, conveys the meaning and causes of the relative condition between ideal (desired) flow features and the corresponding real (actual) flow features for a given WWTU. This has been the approach taken in this book, even though we recognise that the term 'hydraulic efficiency' most probably will not cease to be used in the literature. We argue, however, that its inappropriateness increases further as other flow-related aspects, besides the residence time of individual fluid parcels, have an important influence on the overall treatment efficiency of a WWTU. This occurs, for example, with the effect of turbulent shearing on floc formation and break-up in flocculation units (e.g. Sartori, 2015).

One way to assess hydrodynamic efficiency is from the viewpoint of residence time effect on process efficiency. This usually involves comparing the RTD curve obtained for the real flow pattern in a unit with the idealised flow pattern associated with the treatment process for which the unit was designed. For instance, in continuous flow equalisation units the purpose is to achieve homogeneous properties (e.g. temperature and concentration of some chemical or biological element) in its entire working volume. Thus, the unit should have hydrodynamic characteristics more closely resembling those of the complete mixing flow pattern. If the RTD curves of Figures 2.5 and 2.6 were representative of two wastewater treatment unit setups, then this analysis would favour the adoption of the unit represented by the curves shown in Figure 2.6 as the most adequate to operate as an equalisation unit, admitting that all other design parameters are also met. This type of assessment is further explained and used in the following section, and is the primary assessment approach covered in this book.

2.3 WWTU HYDRODYNAMIC PERFORMANCE ASSESSMENT METHODS

As seen in the previous chapter, the adoption of the flow patterns plug flow and complete mixing has been a usual practice, although, as mentioned in Section 2.2, they are not verified in practice. Consequently, the resulting performance levels of treatment processes in reactors designed under such considerations may be significantly lower than anticipated at the design stage. Compensations to this problem typically include measures such as increasing reagent dosage and building new and different treatment units, which have unnecessary financial implications and can bring other unintended deleterious consequences, such as the formation of health-hazardous disinfection by-products in chlorination units.

Several factors may contribute to the observed deviations between real and ideal flow patters in WWTU, such as flow volumes occupied by recirculating and dead zones, as well as turbulence levels, short-circuiting intensities, and mixing degrees, as defined earlier in this chapter. Thus, knowledge of the hydrodynamic efficiency of a WWTU enables assessment of the degree of departure of real flow hydrodynamic features, relative to their ideal flow counterparts, for a given unit.

In WWTU hydrodynamic efficiency studies, flow pattern characterisation can lead to knowledge on the intensity of recirculating flows and dead zones, short-circuiting, and solute transport and mixing, which serve as a basis for mitigating inefficiencies and improving the hydrodynamic performance. Such characterisation can be done using direct or indirect methods, in both physical and computational settings. Direct assessment methods of the hydrodynamic efficiency typically focus on flow field features, such as mean velocities and/or turbulence components, through measurement, visualisation or estimation with a computational model. Indirect assessment methods of the hydrodynamic efficiency have traditionally been based on Residence Time Distribution curves and associated Hydrodynamic Efficiency Indicators, which can also be obtained through measurements or computer model simulations of tracer experiments. Each assessment approach is described below.

2.3.1 Assessing WWTU Hydrodynamic Performance Using Flow Fields

Two approaches are commonly used in this case: flow visualisation techniques and direct flow field velocity measurements. There are several flow visualisation methods, which normally have the advantages of being non-invasive (i.e. not interfering with flow dynamics, which can occur when measurement devices are inserted in the flow) and allowing for direct visualisation of the flow pattern, including for example recirculating flow zones, advective paths and dispersion layers, which is not possible by means of point measurements. Typical examples of these methods include the injection of dye tracers and/or the release of drift buoys to visualise or infer on the corresponding fluid flows. Flow field assessment is highly recommended in situations requiring detailed diagnostics of WWTU hydrodynamic behaviour, as this approach allows for corrective localised measures to be taken to improve the hydrodynamic efficiency (e.g. Lyn and Rodi, 1990; Teixeira, 1993, 1995b).

Adopting a flow field measurement approach can be impaired by a relatively small amount of suitable, available and cost-effective devices for carrying out such measurements, both in field and in laboratory applications. In field scale units, accessing the interior of treatment units to setup instruments and conduct measurements can be difficult and limit control over certain variables, such as wind-induced shear and currents and any temporal variation of the inflow characteristics associated with diurnal cycles of wastewater generation or supply water demand, for example. Thus, laboratory studies with scaled down units have been a widespread practice in WWTU research studies, allowing for controlled investigation of the flow pattern under a range of operating conditions of interest. Model WWTU design follows established principles and techniques of hydraulic engineering and modelling, such as applying an appropriate scaling law for promoting dynamic similitude between the physical model and the field scale unit when undertaking experiments. For instance, for WWTU that operate with free surface flows, it is common practice to impose Froude number (Fr) similitude – which recognises the key role played by gravitational forces in driving the flow, in both the physical model and the field scale unit – to the detriment of other scaling laws, such as Reynolds number (Re) similitude – which instead focuses on reproducing the relative importance of viscous effects in the flow condition in, for example, pipe reactors. To prevent a non-turbulent flow from occurring in the physical model – which tends to significantly alter key WWTU hydrodynamic features in relation to the field scale unit (Rauen, 2001; Teixeira and Rauen, 2014) – a minimum geometric scale factor (λ) is defined and used when designing the model unit. The definition of this minimum λ value also typically involves considerations related to available laboratory space and the required hydraulic feed system – reservoirs, pumps, piping etc.

Physical models of WWTU are usually undistorted, i.e. the same λ value is used in all three coordinate directions, which confers geometric similarity between the full scale and model units. Thus, surface areas are scaled in proportion to λ^2, while volumes are scaled in proportion to λ^3. By adopting Froudian similitude law to design the physical model experiments,[5] the flow rate

---

5 See Teixeira and Rauen (2014) for a discussion on the implications of this practice in the context of undertaking physical model experiments of chlorine contact units.

is scaled in proportion to $\lambda^{2.5}$, while the hydraulic residence time is scaled in proportion to $\lambda^{0.5}$. Take, for example, a field scale WWTU that is 50 m long by 10 m wide by 4.0 m deep (assumed as the depth of flow) that operates with a design flow rate of 1.0 $m^3.s^{-1}$. If scaled-down by 10 times, it is reproduced in the laboratory by a 5.0 m x 1.0 m x 0.4 m physical model. The cross-sectional flow area (assuming that the 50 m long dimension is in line with the streamwise flow direction, from inlet to outlet) reduces from 40 m^2 in the field scale unit to 0.4 m^2 in the physical model, while the corresponding fluid volume reduces from 2,000 m^3 to 2.0 m^3. The flow rate in the physical model is approximately 320 times lower than in the field scale unit, i.e. it reduces from 1.0 $m^3.s^{-1}$ to circa 3.2 $L.s^{-1}$. The hydraulic residence times is approximately 3.2 times lower, reducing from around 33 minutes in the field scale unit to just under 11 minutes in the physical model. Care should be taken, at the design stage, to minimise the occurrence of substantial so-called scale effects – the overall impact, on the flow field, of incomplete dynamic similitude in terms of all other neglected hydro-dynamic forces, such as surface tension, compression, pressure gradients etc. Another aspect to note is that such reduction in residence time can have implications when simulating the treatment process in the laboratory model study. This is because a) most treatment processes have a kinetic (time-dependent) component associated with the physical, chemical and/or biological reactions involved in so-called treatment, and b) related particles, flocs, organisms and/or reaction rates are not scaled-down together with the WWTU structure and flow conditions. Process-specific attention needs to be paid to mitigate such issues in addressing WWTU research needs using physical modelling techniques, particularly as some processes also have minimum requirements, for example, in terms of the hydraulic residence time in a WWTU (as shown for anaerobic reactors by Chernicharo, 1997).

There are many examples of flow field visualisation using dye tracers in the international literature, and only a small amount of publications reporting comprehensive flow field velocity measurements in WWTU. Studies such as those outlined in this section highlight the historical importance of physical modelling techniques in improving knowledge and descriptions of hydro-dynamic processes in WWTU, as well as their effects on treatment process efficiency.

For instance, Hart et al. (1975) injected small amounts of black ink at three depths (top, centre and bottom of the water column) and in various positions in a reduced scale reactor model. Based on the streamlines traced in the flow by the dye, it was possible to identify key regions of recirculating flow, short-circuiting and dead zones. The analysis allowed for detecting flow conditions, which were closer to plug flow, among several flow patterns caused by modifications implemented to the reactor model. As the flow visualisation method is essentially qualitative, it is usually applied together with at least one of the other flow characterisation methods discussed herein.

The analysis and comparison of flow field measurements complemented by visual inspection were used by Falconer and Tebbutt (1986) to define which modification to a scale model would lead to flow conditions closer to plug flow. A solution of potassium permanganate was injected at the unit inlet under various model modifications, allowing for observation of the occurrence of pronounced recirculating flow and dead zones in each case, leading to the definition of which setup had closer-to-ideal flow conditions.

Teixeira (1993) undertook point velocity measurements using Laser Doppler Anemometry (LDA) in the laboratory WWTU model illustrated in Figures 2.7, 2.8 and 2.9. It consisted of a 1:4 scaled-down model of a chlorine contact unit[6] located in England. Baffles were used to provide an eight compartment serpentine flow configuration. An open channel was used as the inlet device and a sharp-crested rectangular weir was the outlet device. Both devices spanned the width of the corresponding compartment. At the inlet channel there was also a partly submerged sharp-crested weir, as can be seen in Figures 2.8 and 2.9.

Each point velocity measurement lasted for 2 min or 15 min and was conducted at a frequency of 50 or 100 kHz, depending on the mean flow speed (relatively high and low respectively), as required for an unbiased determination of each parameter (mean velocity components and turbulence intensities respectively). After each time series record was obtained, the device was traversed longitudinally, laterally or vertically by a few centimetres to a neighbouring point for another measurement, and the procedure was repeated until a certain flow area of interest was fully characterised. After measuring several flow areas, Teixeira (1993) obtained a detailed and high resolution characterisation of the velocity and turbulence fields in the model unit.

Contrary to the ideal flow condition for a contact unit – plug flow with low turbulence levels – Teixeira (1993) found a complex flow field in the model unit, with recirculating flow zones and relatively high turbulence levels in various regions. As usual in serpentine flow configurations, horizontal recirculation zones were detected at the lee side of baffle tips, and the location of the corresponding flow separation and reattachment points were identified based on the velocity fields, as shown in Figure 2.10.

---

6 This type of WWTU is typically used for chlorine disinfection of water and wastewater. In such units, the occurrence of excessively long residence times of fluid particles trapped in recirculation zones and dead zones is often associated with enhanced formation of potentially hazardous by-products, which is an undesirable effect of non-ideal flow features. Such zones are typically formed in corners and downstream of baffle lees, and originate in boundary layer separation owing to adverse pressure gradients. Their presence can often be diagnosed using RTDs and their effect is essentially on the residence time of fluid portions.

FIGURE 2.7 Scaled-down chlorine contact unit (Tank 2) used in laboratory experiments by Teixeira (1993); (Shiono and Teixeira, 2000)[7]

As shown in Figure 2.11, in the first compartment, due to the fact that the inflow entered the unit with a mean depth of only around 20% of the compartment flow depth, a plane wall jet-like advective zone was formed near the free surface. Upon reaching the opposite wall, such relatively high speed flow (compared to the cross-section mean velocity) was deflected mainly downwards towards the bed and sideways towards the second compartment. The flow portion that remained in the first compartment was again deflected by the bed and flowed in the counter-streamwise direction (reversed flow) along the bed towards the inlet wall. Upon reaching the inlet wall another deflection occurred, this time upwards towards the inlet section. The loop-like trajectory that

7 Turbulent characteristics in a baffled contact tank, Shiono & Teixeira, *Journal of Hydraulic Research*, 38 (6), 403–416, 2000, © International Association for Hydro-Environment Engineering and Research, reprinted by permission of Informa UK Limited, trading as Taylor & Francis Group, www.tandfonline.com on behalf of International Association for Hydro-Environment Engineering and Research.

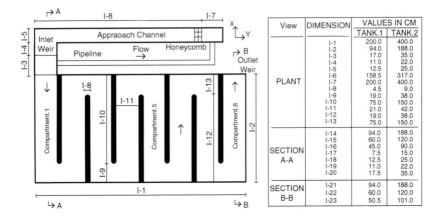

View	DIMENSION	VALUES IN CM	
		TANK.1	TANK.2
PLANT	I-1	200.0	400.0
	I-2	94.0	188.0
	I-3	17.0	35.0
	I-4	11.0	22.0
	I-5	12.5	25.0
	I-6	158.5	317.0
	I-7	200.0	400.0
	I-8	4.5	9.0
	I-9	19.0	38.0
	I-10	75.0	150.0
	I-11	21.0	42.0
	I-12	19.0	38.0
	I-13	75.0	150.0
SECTION A-A	I-14	94.0	188.0
	I-15	60.0	120.0
	I-16	45.0	90.0
	I-17	7.5	15.0
	I-18	12.5	25.0
	I-19	11.0	22.0
	I-20	17.5	35.0
SECTION B-B	I-21	94.0	188.0
	I-22	60.0	120.0
	I-23	50.5	101.0

FIGURE 2.8   Plant view and dimensions of the model unit assessed by Teixeira (1993); (Shiono and Teixeira, 2000)[8]

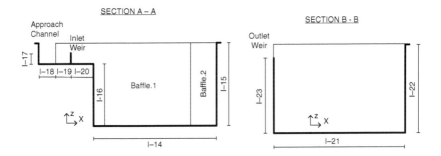

FIGURE 2.9   Side views (as indicated in Figure 2.8) of the model unit assessed by Teixeira (1993); (Shiono and Teixeira, 2000)[9]

8   Turbulent characteristics in a baffled contact tank, Shiono & Teixeira, *Journal of Hydraulic Research*, 38 (6), 403–416, 2000, © International Association for Hydro-Environment Engineering and Research, reprinted by permission of Informa UK Limited, trading as Taylor & Francis Group, www.tandfonline.com on behalf of International Association for Hydro-Environment Engineering and Research.

9   Turbulent characteristics in a baffled contact tank, Shiono & Teixeira, *Journal of Hydraulic Research*, 38 (6), 403–416, 2000, © International Association for Hydro-Environment Engineering and Research, reprinted by permission of Informa UK Limited, trading as Taylor & Francis Group, www.tandfonline.com on behalf of International Association for Hydro-Environment Engineering and Research.

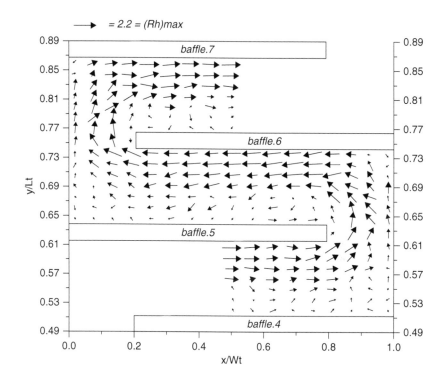

FIGURE 2.10   Mean velocity field measured at the mid-depth plane of the flow region from the fifth to the seventh compartments of the model unit assessed by Teixeira (1993). Dimensions were normalised by the overall internal width (Wt) and length (Lt) of the model unit. At the top of the figure the reference velocity value was normalised by the mean cross-sectional velocity of 1.04 cm.s^{-1} (Shiono and Teixeira, 2000)[10]

characterises recirculation zones was closed where such an upwards current met with the incoming flow. Thus, the flow field in the first compartment was dominated by a large recirculation zone in the vertical plane.

10 Turbulent characteristics in a baffled contact tank, Shiono & Teixeira, *Journal of Hydraulic Research*, 38 (6), 403–416, 2000, © International Association for Hydro-Environment Engineering and Research, reprinted by permission of Informa UK Limited, trading as Taylor & Francis Group, www.tandfonline.com on behalf of International Association for Hydro-Environment Engineering and Research.

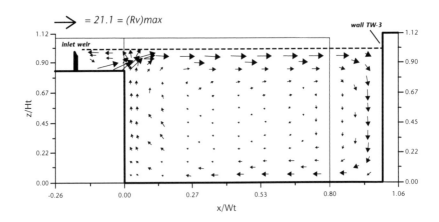

FIGURE 2.11   Mean velocity field measured at the mid-width plane of the first compartment of the model unit assessed by Teixeira (1993). Dimensions were normalised by the overall internal width of the model unit (Wt) and water depth (Ht). At the top of the figure the reference velocity value was normalised by the mean cross-sectional velocity of 1.04 cm.s^{-1} (Shiono and Teixeira, 2000)[11]

The second compartment was also dominated by a large recirculation zone in the vertical plane, as shown in Figure 2.12. The deflected flow coming from the first compartment at wall TW-3 travelled near the bed in the streamwise direction towards wall TW-1, where it was deflected mainly upwards towards the free surface and sideways towards the third compartment. Upon reaching the free surface this portion of flow was deflected towards wall TW-3, forming a reversed flow structure along the free surface. Upon meeting other currents from the baffles and from the first compartment (not shown in the figure), another loop-like trajectory was established, which occupied around 60% of the vertical mid-width plane area of the second compartment.

Further measurements undertaken by Teixeira (1993) in the rest of the model unit showed that the flow pattern was negatively affected by the inlet configuration up to the fourth compartment. Thus, such a direct characterisation of the velocity field showed that the defining feature of the plug flow pattern – all fluid parcels having equal velocity and following parallel and straight paths –

11 Turbulent characteristics in a baffled contact tank, Shiono & Teixeira, *Journal of Hydraulic Research*, 38 (6), 403–416, 2000, © International Association for Hydro-Environment Engineering and Research, reprinted by permission of Informa UK Limited, trading as Taylor & Francis Group, www.tandfonline.com on behalf of International Association for Hydro-Environment Engineering and Research.

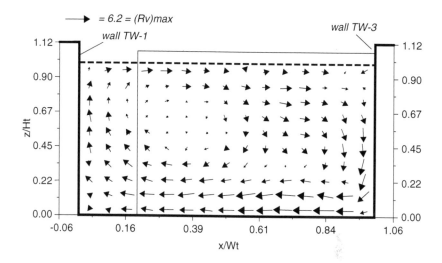

FIGURE 2.12 Mean velocity field measured at the mid-width plane of the second compartment of the model unit assessed by Teixeira (1993), with normalisation applied as in Figure 2.11 (Teixeira, 1993)

only occurred from the fifth to the eighth compartments, apart from the regions occupied by horizontal recirculation zones such as illustrated in Figure 2.10.

Rauen (2005) undertook a hydrodynamic efficiency assessment of another physical model of contact unit using acoustic Doppler velocimetry, tracer experiments and CFD modelling. Among the setups thus assessed was one (illustrated in Figure 2.13) designed to mimic Teixeira's (1993) eight-compartment baffled unit with a channel-like inlet device and a sharp-crested weir as the outlet device, but with dimensions: 2.0 m long (in the inflow streamwise direction), 3.0 m wide (where each compartment was 0.36 m wide) and a flow depth of 1.0 m in the unit.[12] Key flow field features reported by Teixeira (1993) (as outlined in Figures 2.10, 2.11 and 2.12) were generally reproduced in this unit setup, including the extent of significant three-dimensional flow.

The distribution of turbulence parameters in the unit was investigated by Rauen (2005) using CFD and led to results such as those shown in Figure 2.14 for the eddy viscosity. It can be noted in Figure 2.14 that the eddy viscosity peaked in regions of high shear, as caused by flow reattachment and flow

---

12 Other baffling arrangements and a pipe inlet device were also tested by Rauen (2005) for their effect on the hydrodynamic efficiency of the unit, as discussed in Chapter 3.

FIGURE 2.13   Scaled-down chlorine contact unit used in laboratory experiments by Rauen (2005) (Rauen et al., 2008)[13]

deflection by walls, baffles and the incoming flow in compartment one. Near-bed eddy viscosity values were generally lower, by an order of magnitude, than elsewhere in the flow, where results were rather uniform apart from the maxima regions mentioned above. Isocontour plots, such as these, can aid in 'big picture' analysis of parameter distribution and quick identification of remarkable flow features in WWTU hydrodynamic assessments.

Complex three-dimensional flow, reversed flow structures or low flow velocities observed in laboratory studies can limit the application of cost-effective flow measurement devices, such as propeller meters (Stamou and Rodi, 1984; Teixeira, 1993). High cost is still an issue with the most advanced systems, which include for example, laser or acoustic velocimetry (Teixeira, 1993; Rauen, 2005), but their usage can be justified particularly where even small improvements in hydrodynamic performance lead to substantial economic gains, as highlighted in the Preface. More recently, particle-induced velocimetry (PIV) has become a popular, non-invasive technique for velocity measurements, and relatively low cost devices (of the order of thousands of US dollars) can be implemented and used to good purpose.

---

13 Reprinted with permission of Elsevier, from *Chemical Engineering Journal*, 137, 550–560, Rauen, Lin, Falconer, Teixeira, CFD and experimental model studies for water disinfection tanks with low Reynolds number flows, 2008.

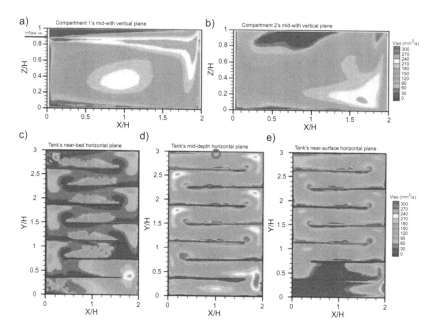

FIGURE 2.14 Eddy viscosity (*Visc*) distributions obtained using CFD for the eight-compartment model unit with channel inlet, where the circle in figure d represents an area assessed in detail for other turbulence parameters (not included herein) (dimensions normalised by $H = 1.0$ m) (Rauen, 2005)

Some of the above difficulties in measuring WWTU flow fields and obtaining process-meaningful results in physical models might be circumvented through the use of a computational modelling approach. The purpose, in this case, is to firstly simulate, as accurately as possible, WWTU hydrodynamics to estimate, for example, mean velocity distributions and turbulence levels in different parts of a unit (e.g. Rauen et al., 2012). Such an approach is usually performed with the aid of a CFD model, and typically requires:

- A suitable mathematical framework (governing equations and boundary conditions) to adequately represent key flow processes in a given unit. Usually this means that at least a two-dimensional mathematical framework should be used, with a fully three-dimensional approach often being the most suitable to simulate all but those units in which the flow pattern approaches plug flow.
- Suitable numerical methods to enable computational processing and resolution of the systems of equations.

- Suitable interface and user experience to enable the preparation and execution of simulations.
- Suitable data for model calibration and validation, prior to achieving trustworthy results for a given unit and flow condition of interest.

Even with the current availability of highly sophisticated CFD models, experiments conducted at the laboratory and/or prototype scales will arguably continue to be required, prior to full-scale WWTU implementation, at least to provide data for model calibration and validation. Such data requirements tend to increase with problem complexity, and particularly for multiphase flows when chemical and biological processes are also simulated and/or interfere with the physical processes.

A CFD based study of WWTU may involve solely hydrodynamic modelling, aiming to investigate the level of proximity between the real (simulated) and the ideal flow conditions (e.g. Sartori et al., 2015). However, a common subsequent step of interest in CFD modelling of WWTU treatment processes can be to simulate the transport of solutes and/or particles as they flow through the unit. These processes can range from the relatively simple case of a mass conservative solute flowing through a clean water unit, to the much more complex problem of a reacting three-phase (gas–liquid–solid) flow in a sewage treatment unit. Problem complexity grows with the number of variables that control the efficiency of the treatment process in a given scenario. Any by-product formation and interacting effects among substances might also have to be included (e.g. Zhang et al., 2000; Angeloudis et al., 2014a, 2014b), as well as any known dependency upon concentrations, temperature, pH, etc. A realistic representation of such processes by CFD models depends upon the availability of a suitable mathematical framework to describe and simulate the relevant physical, chemical and/or biological processes. In turn, this requires that the related basic scientific foundations have been laid to a considerable extent, and/or that CFD modelling is used to support the advancement of an existing mathematical framework, as suggested by Laurent et al. (2014).

2.3.2 Assessing WWTU Hydrodynamic Performance Using Residence Time Distribution Curves and Hydrodynamic Efficiency Indicators

WWTU hydrodynamic efficiency assessment methods based on tracer techniques are the most commonly used, as applied in the field and laboratory scales through physical experimentation, or via computational simulations of hydrodynamic and solute mixing processes that mimic a tracer experiment.

Tracers (substances or devices) simulate the movement of one or more fluid parcels, including salts, fluorescent substances and floating devices, citing sodium chloride, rhodamine and styrofoam, respectively, as examples used in hydrodynamic studies. Broad guidance on the use of tracers in these types of studies can be found in Tchobanoglous et al. (2003) and in Rigo and Teixeira (1995). These latter two authors present, for example, a set of aspects to consider when choosing a tracer, as follows:

- Level of similarity in terms of characteristic to what is to be traced (density, solubility in water, etc.).
- Adsorption level to suspended particles, biomass and walls, etc.
- Susceptibility to be degraded by biomass.
- Level of substance conservation (mass, flourescence, etc.) over time.
- Cost and availability.
- Safety and easy of use.

Tracer injection can be either instantaneous or continuous, while monitoring typically involves one of the following options (for tracer substances that meet the above requirements):

- Manual collection of fluid samples at the specified monitoring location and at certain time intervals for subsequent laboratory processing; the associated temporal sampling resolution is usually much lower than for the following alternatives.
- Using an automated system of pumping and circulation through a portable device used to read the tracer property of interest.
- Via direct measurement of the tracer property of interest with a watertight device inserted in the flow field without causing significant disturbance.

Tracer tests in WWTU reported in the international literature have mostly focused on determining the RTD or passage curve at the outlet of a unit under certain operating conditions, and following an instantaneous injection of a known/controlled quantity of tracer at the inlet section. Outlet concentration monitoring can be undertaken using automated methods (e.g. Teixeira et al., 2002) or by manual sampling. In order to satisfy the principle of mass conservation during experimentation with tracers, and thus allow the proper use of several supporting procedures for analysis of results, their properties should remain practically unchanged during this period. This type of application confers a unit-wide, 'bulk' response of the effect of all flow mixing processes taking place in the assessed unit. No detailed information is gathered on the longitudinal, lateral or vertical variations of variables such as the diffusion coefficient and substance concentration using this type of test. Variants of such an approach involve monitoring tracer concentration inside a WWTU to obtain more detailed information on solute transport and mixing processes. However, depending on flow complexity, analyses of such curves will only be possible with the aid of velocity vector maps (Teixeira and Shiono, 1994).

RTD curves obtained in a given unit can be compared with their idealised flow counterparts – plug flow and complete mixing, as exemplified in Section 2.2 – to enable a qualitative assessment of WWTU hydrodynamic performance. RTD curves obtained in different units or in different geometric and/or operational configurations of the same unit can be compared to infer the effect of modifications on WWTU hydrodynamic performance, aiming to optimise their design.

From the RTD a number of Hydrodynamic Efficiency Indicators (HEI) can be derived to enable semi-quantitative analyses of WWTU hydrodynamic performance, which complement the RTD-based assessment. Examples of such types of assessments are shown later in this chapter. Here we present a basis for the analysis methods and definition of key components and parameters.

2.3.2.1 Hydrodynamic Efficiency Indicators (HEI)

Indicators of short-circuiting intensity and degree of mixing, extracted from the unit's passage curves, have been shown to be very appropriate in the evaluation of hydrodynamic performance of WWTU. This has been achieved using parameters extracted from the passage curves as indicators of the level of short-circuiting and mixing in WWTU.

The relative abundance of HEI in the international literature, both for short-circuit and for mixing, can sometimes make it difficult to analyse and interpret results, and also to compare results obtained in different scenarios. An ideal indicator adequately represents the phenomenon or feature to which it is related, is robust enough so that its indicating ability does not suffer biases associated with changing conditions in different applications and, ideally, can be meaningfully used in a wider context – such as explicitly figuring in a mathematical framework that describes processes of interest.

Thus, Siqueira (1998) and Teixeira and Siqueira (2008) assessed the behaviour of a selection of 14 HEI found in the international literature (eight for short-circuiting and six for mixing) for their physical basis and statistical reproducibility. The aim was to identify the most suitable HEI to assess short-circuit and mixing in WWTU, based on a joint interpretation of both physical processes. A range of flow patterns in distinct WWTU physical models were tested to assess whether the robustness of HEI responses depended on the levels of short-circuit and mixing. Such dependence was seen as a disadvantageous bias for their usage, as the flow pattern in a given WWTU under assessment may not be known *a priori* and may vary due to the implementation of flow modifier structures and/or variation of operating conditions. Key HEI thus tested are described below, with the aid of Figures 2.15 and 2.16 for their graphical representation.

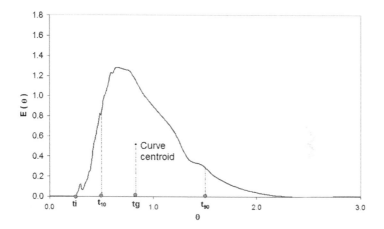

FIGURE 2.15 Graphical representation of indicators t_i, t_{10}, t_g and t_{90} on $E(\theta)$ curve associated with a hypothetical real flow pattern (adapted from Siqueira, 1998)

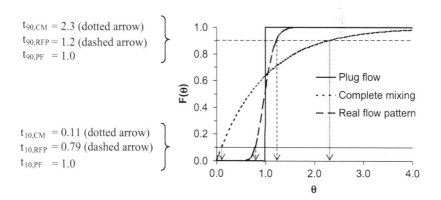

FIGURE 2.16 Cumulative passage curves for a hypothetical real flow pattern (RFP) and the idealised plug flow (PF) and complete mixing (CM) flow patterns, with arrows identifying the corresponding t_{10} and t_{90} values on the θ axis

Initial arrival time of tracer at monitoring location (t_i): this HEI is the one most closely related to the predominantly advective short-circuiting phenomenon, because it is associated with streamlines with displacement of fluid parcels with shorter residence times in the reactor.

Due mainly to the possibility of experimental errors, t_i determined by physical modeling is subject to high statistical reproducibility, which can be increased in case of monitoring unit passage curves of low temporal resolution as well as non-automated.[14]

Time of passage of 10% of the tracer mass through the reactor outlet section (t_{10}): by delimiting a small portion of the initial part of the passage curve, this indicator, similar to t_i, is a good representation of the short-circuiting physical phenomenon. In situations of obtaining t_{10} through physical modeling, because it is related to a portion of area and not to a point of the reactor passage curve, it tends to present higher statistical reproducibility than t_i. Figure 2.15 identifies t_{10} for a hypothetical real flow pattern using the $E(\theta)$ curve, while Figure 2.16 illustrates the determination of such an indicator using the $F(\theta)$ curve for a hypothetical real flow pattern and for the idealised plug flow and complete mixing flow patterns.

Dispersion index (σ^2): Based on its definition given by Equation 2.10, this index represents the tracer mass scattering level in the reactor, expressed as the variance of the corresponding passage curve. Thus, conceptually, it is a good indicator of mixing.

$$\sigma^2 = \frac{\sigma_t^2}{t_g^2} \qquad (2.10)$$

where σ_t^2 is the variance of the normalised tracer passage curve, calculated using Equation (2.11), and t_g is the time associated with the centroid of the area under the RTD curve (indicated in Figure 2.15 and calculated in normalised form using Equation (2.12) as a weighed time average, with the product $[E(\theta).d\theta]$ serving as the weigh for each θ value in the time series).

$$\sigma_t^2 = \frac{\int_0^\infty \theta^2.E(\theta).d\theta}{\int_0^\infty E(\theta).d\theta} - t_g^2 \qquad (2.11)$$

---

14 Statistical variability was assessed as a measure of variation of indicator values relative to an average result calculated with such values. Thus, the lower the statistical variability, the better the reproducibility of that indicator, which can be associated with indicator robustness.

$$t_g = \frac{\int_0^\infty \theta.E(\theta).d\theta}{\int_0^\infty E(\theta).d\theta} \tag{2.12}$$

The determination of σ^2 by Equation 2.10 requires the use of the entire unit's curve of passage, unlike the great majority of the mixing indicators, whose determination is based on only some of its points. In addition to considering the entire passage curve in determining its value, and thus, globally, all geometrical aspects and of fluid dynamic which define the mixing degree in the unit, another reason for σ^2 being one of the most reported mixing indicators in the literature is its relation with the global mixture coefficient (d), which was presented in Section 2.1, highlighting its use in the practical application of hydro-kinetic models, as exemplified in Chapter 4. For closed basins, as defined by Levenspiel (1999) – the case of most typical WWTU – the relationship between σ^2 and d is as follows:

$$\sigma^2 = 2d - 2d^2 \left(1 - e^{-d^{-1}}\right) \tag{2.13}$$

Morrill index (Mo): this index is calculated as the ratio between the time for 90% of the injected tracer mass leave the reactor (t_{90} as illustrated in Figures 2.15 and 2.16) and t_{10} ($Mo = t_{90}/t_{10}$). The occurrence of recirculating flow and dead zones in portions of a WWTU contribute to increasing this indicator, as this tends to provoke long RTD tails and thus, on one hand, higher t_{90} values, and, on the other, lower t_{10} values (as it is known, there exist, usually, a positive correlation between the volume occupied by recirculating flows and dead zones in a WWTU and the intensity of short circuit; thus, higher t_{90} values are typically associated with lower t_{10} values in a unit). A relatively long RTD tail reflects the fact that tracer mass is trapped in such zones and then gradually released into the main flow path through momentum and mass exchange, usually caused by shear at the interface with preferential flow paths (the main advective flow through the unit).

In summary, based on the assessment of Siqueira (1998), Siqueira et al. (1999) and Teixeira and Siqueira (2008), the following observations can be drawn:

- Considering the eight short-circuiting indicators assessed, only t_{10} fully met the requirements considered by the authors for an appropriate short-circuit indicator: it is directly associated with the early portion of the passage curve and has high statistical reproducibility, being unbiased in relation to the type of flow pattern. It is also used in hydro-kinetic models, as seen in Section 1.2.

- t_i provides a relatively quick diagnostic of the flow pattern in the unit, as it does not require the determination of the whole RTD curve. Even though this indicator had higher statistical variability than t_{10}, it was deemed robust enough for use in engineering problems. It is, thus, a valuable short-circuit indicator with practical flow pattern diagnostic relevance.
- Thus, assessments of the short-circuit intensity in WWTU should preferably rely on t_{10} and t_i.
- Mo was found to have low statistical variability, to be unbiased by the type of flow pattern and to adequately represent the mixing levels in a flow pattern. The only drawback of this indicator is its absence from hydro-kinetic models, as far as the authors are aware, which prevents linking WWTU hydrodynamic efficiency assessments with treatment efficiency estimates.
- σ^2 has a sound theoretical foundation as a mixing indicator due to taking into account the whole passage curve. It is used widely, in particular in assessments involving determination of the dimensionless dispersion coefficient (d) for subsequent hydro-kinetic model use. However, for flow patterns tending to plug flow, this indicator presents low values (close to zero), which can lead to relatively large proportional discrepancies between assessments (even replicates) and lead to misleading conclusions.
- Thus, for hydrodynamic efficiency assessments only in WWTU, Mo is the most recommended indicator, but in situations when a WWTU hydrodynamic efficiency assessment is accompanied by any process efficiency estimation, such as using a hydro-kinetic model that relies on d (when available), then the procedure can involve the joint use of Mo and σ^2.

2.3.2.2 HEI values associated with the idealised flow patterns

Table 2.1 shows the values[15] of t_i, t_{10}, Mo and σ^2 associated with the plug flow and complete mixing flow patterns. These analytically determined values can be taken as reference points in comparisons with the corresponding results obtained from RTD curves of real flow patterns, in the process of assessing the hydrodynamic efficiency of a WWTU.

Experience shows that plug flow results depart substantially from those observed in real flow patterns, even in situations with low mixing levels.

---

15 Time values for t_i and t_{10} are considered in terms of normalised time so that they represent values on the Θ axis.

TABLE 2.1 Values of Hydrodynamic Efficiency Indicators (HEI) for idealised flow patterns

HEI	PLUG FLOW	COMPLETE MIXING
t_i	1.0	0.0
t_{10}	1.0	0.11
σ^2	0.0	1.0
Mo	1.0	21.8

A more realistic target scenario for hydrodynamic optimisation studies can be created using a theoretical RTD curve, such as Equation 1.5, assuming baseline mixing effects caused by dispersion only, i.e. $d = d_L$, with d_L calculated using Equation 2.3 (Falconer and Tebbutt, 1986). As required to estimate D_L in Equations such as 2.1 and 2.2, a rough estimate for the friction velocity can be obtained based on the assumption of 1D uniform flow, as:

$$u_* = \sqrt{\frac{\tau_0}{\rho}} = \sqrt{gRS} \tag{2.14}$$

where τ_0 = bed shear stress; ρ = fluid density; g = acceleration due to gravity; R = hydraulic radius; S = energy line gradient, which for uniform flow can be obtained from Manning's equation as:

$$S = \frac{U^2 n^2}{R^{\frac{4}{3}}} \tag{2.15}$$

where U = cross-sectional mean flow velocity; n = Manning's roughness coefficient.

Such a procedure enables a 1D-type approach for comparing hydrodynamic performance of WWTU using measured and expected tracer test results, where plug flow is a relevant flow pattern.[16]

---

16 Care should be taken when using such an approach for determining d, as it is recommended for scenarios in which the WWTU flow condition is comparable to that in a straight open channel and, in practical engineering terms, fluvial/open channel hydraulics equations are applicable. Where such considerations are at first inappropriate, then it is possible that altering the WWTU geometry such that the flow condition approaches the above scenario will favour plug flow and enable the use of such a type of model.

An example to illustrate the importance of taking mixing effects into account to obtain an idealised RTD curve for a WWTU was obtained by Almeida (1997), as shown in Figure 2.17. Such a RTD curve provided results for the short-circuiting indicators t_i and t_{10}, as well as the mixing indicators σ^2 and Mo, as shown in Table 2.2.

Table 2.2 shows that plug flow values for t_i and t_{10} are 79% and 22% respectively higher than their counterparts considering dispersion effects. Concurrently, the plug flow value of Mo was 36% lower than the more realistic (albeit estimated) plug flow with dispersion condition, while it would not be appropriate to infer on σ^2 in terms of percentage variation due to the fact that it is zero under the pure plug flow condition. These substantial differences obtained for t_i, t_{10} and Mo guide us to the adoption, whenever possible, of plug flow with dispersion as the reference flow pattern (for being a more realistic target) for hydrodynamic optimisation studies where plug flow is the theoretical idealised flow pattern of interest.

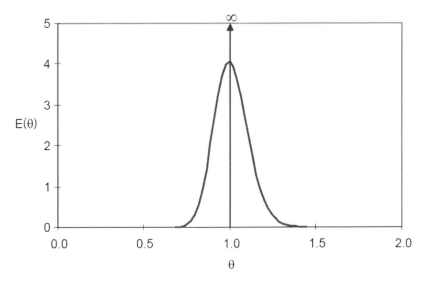

FIGURE 2.17   RTD curves for idealised plug flow (vertical line) and 'plug flow with dispersion' (bell-shaped line) patterns ($d = d_L = 0.0144$) (adapted from Almeida, 1997)

TABLE 2.2 Theoretical values of HEI for a WWTU investigated by Almeida (1997)

HEI	PLUG FLOW	PLUG FLOW WITH DISPERSION
t_i	1.0	0.56
t_{10}	1.0	0.82
σ^2	0.0	0.03
Mo	1.0	1.56

2.3.2.3 Internal Assessment of WWTU Flow Patterns Using Tracer Experiments

As mentioned earlier, a less traditional type of tracer experimentation in WWTU involves injection and/or monitoring at locations other than the inlet and outlet sections of a WWTU. In principle, these two actions can be performed in virtually any accessible location within the flow field of a WWTU. Outputs such as RTD curves can be used to infer on the combined effect of all transport and mixing processes taking place between the injection point(s) and the monitoring location(s).

For instance, it may be of interest to partition a hydrodynamic efficiency assessment between the inlet section (injection) and certain locations of interest within the unit (multiple monitoring points). Such a monitoring approach was undertaken by Teixeira (1993) to assess how the mixing levels varied longitudinally, laterally and vertically in the laboratory model of a baffled chlorine contact unit illustrated in Figure 2.7. The longitudinal variation was assessed based on the shape of the RTD curves obtained in the fifth, sixth and seventh compartments, as well as at the outlet section of the model unit, as shown in Figure 2.18. It can be seen that the RTD curves gradually become shorter and wider in the downstream direction, as a result of mixing processes along the flow.

Results obtained in such tracer tests were analysed together with measured mean velocity and turbulence intensity fields, and allowed for understanding and quantifying the extent of impact of the inlet device on the mixing levels in the unit. It was found, for instance, that approximately the first half of the serpentine flow length was affected by three dimensional flow features owing to such an inlet device, with knock on impacts on, for example, the mixing coefficient.

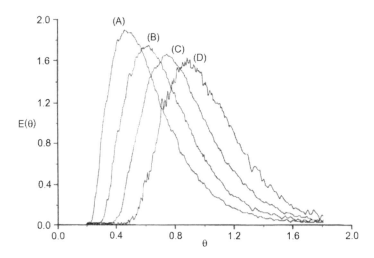

FIGURE 2.18   RTD curves obtained in the fifth compartment (curve A), sixth compartment (curve B) and seventh compartment (curve C), as well as at the outlet section (curve D) of the model unit shown in Figures 2.7–2.9 (adapted from Teixeira, 1993)

Results such as these indicated that:

i)   Care should be taken when designing or changing specific components of a WWTU, such as inlet and outlet devices, baffles and flow deflectors, for their effect (positive or otherwise) on the flow pattern may not be negligible.

ii) There can be a strong interdependence among distinct causes of short-circuit and mixing processes, which has to be taken into account to determine the spatial distribution and relative importance of their impacts on the hydrodynamic efficiency of a WWTU. For instance, Teixeira and Rauen (2014) showed that the plug flow pattern was more responsive than the complete mixing flow pattern to the variation of scale in model units, particularly under low turbulence levels.

Internal flow pattern assessments by means of tracer experimentation were also used by Teixeira et al. (2004) and by Angeloudis et al. (2016) to calibrate and/or validate CFD model predictions. Due to the fact that distinct WWTU can sometimes yield similar RTD curves (Levenspiel, 1999), the higher level of detail obtained using such internal assessments can contribute to reduce uncertainty and improve the robustness of CFD model implementation and usage.

A different type of tracer test uses solid tracers, such as drift buoys, to identify flow trajectories within water bodies. Drogue tracking can be done via Global Positioning System (GPS) in the case of relatively large WWTU, such as full scale ponds (e.g. Barter, 2003), or via image analysis in the case of scaled-down laboratory studies (e.g. Shilton and Bailey, 2006).

Effects of WWTU Setup on Hydrodynamic Performance

3

3.1 INTRODUCTION

The concept of hydrodynamic efficiency is embedded in the design practice of several types of WWTU, even if not explicitly referred to and recognised as such. This chapter exemplifies how hydrodynamic performance of WWTU can be affected by design aspects, with a view to providing a basis for outlining a rational hydrodynamic design and assessment procedure in Chapter 4.

The working volume, or reactive volume V of a WWTU is given by Equation 3.1:

$$V = Q.T \qquad (3.1)$$

where Q is the design flow rate and T is the theoretical residence time (a design input parameter).

With the obtaining of V, the geometry and dimensions of the WWTU can be defined, which usually have as limiting factor the available space. Reactor shape variation, from regular to irregular formats (circular, square, rectangular, triangular, 'L-shaped', among others), is often caused by such space restrictions but is mostly guided by conventional design practice for specific processes.[1] The shape of a WWTU is a key governing factor of its hydrodynamic performance, as the flow processes upon which such performance depends (defined in Chapter 2) occur in the enclosed environment defined by the reactor 'shell' (i.e. walls and bed, in the absence of flow modifier devices[2]), as well as the free surface (if it is not a pressurised vessel).

The performance of some treatment processes is highly dependent upon the depth of the reactive volume, as observed for stabilisation ponds. In such cases, the depth of the working volume of a reactor, or the flow depth, is, similarly to the flow rate and theoretical detention time, a previously defined design parameter. Otherwise, the depth can be adjusted to achieve the desired working volume, i.e. once the plane area for reactor construction is defined, the designer aims for the depth that satisfies the working volume of the reactor.

Three factors will be very important to the definition of the final configuration of WWTU: the designer experience and judgement, as well as on his/her access to the technical-scientific literature in general (design manuals, specialised technical journals, etc.). Thus, complementary design elements[3] can be taken into account, such as flow diffusers, deflectors, baffles, straighteners ('honeycombs', perforated plates, etc.). Figures 3.1 to 3.4 illustrate some of these design elements. This is partly because units of considerable dimensions (i.e. of the order of tens of metres wide and long) are often used, together with inlet devices that consist of open channels or pipes that are narrower, by one or two orders of magnitude, than the wall onto which they are connected. This feature alone generates demand for flow pattern correction measures that are ultimately aimed at improving flow uniformity, in cases where plug flow is the target flow pattern.

Changing a reactor setup can lead to substantial variation of its hydrodynamic performance and, thus, of the treatment process efficiency. Analyses of flow velocity fields, RTDs and HEIs can be used to assess and determine the impact and relative importance of each modification on the flow pattern, relatively to the idealised flow pattern of interest.

---

1 As prescribed in design standards and manuals, as well as found in related technical literature, e.g. Howe et al. (2012), Libanio (2010), Richter (2009), Jordao and Pessoa (2005), Sykes et al. (2003), Tchobanoglous et al. (2003), Keller and Pires (1998), Chernicharo (1997), Stevenson (1997) and Von Sperling (1996a, 1996b).

2 The effect of which is exemplified in this chapter.

3 Factors such as the reduction of reactive volume due to insertion of flow-modifier structures, access to the inner part of reactors for maintenance, construction costs and admissible head losses should be considered with care at the planning stage. For example, there may be minimum depth requirements for the later usage of boats or other devices to remove settled material from sedimentation units.

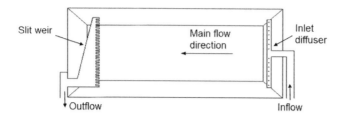

FIGURE 3.1 Top view of a WWTU with inlet diffuser and outlet slit weir plate as examples of complementary design elements (adapted from Almeida, 1997)

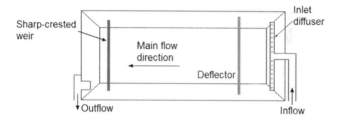

FIGURE 3.2 Top view of a WWTU with inlet diffuser, flow deflector and sharp-crested rectangular weir as examples of complementary design elements (adapted from Almeida, 1997)

3.2 BAFFLING

Baffling is one of the most common WWTU design modifications applied to promote plug flow-like conditions. The primary effect of using baffles in WWTU is to promote a channel-like straight flow and prevent the formation of large recirculation regions,[4] which are associated with the overall intensities of short-circuiting and mixing in the WWTU. In terms of geometrical parameters, baffling is usually carried out in such a way as to reduce flow

---

4 See example given in Figure 2.10.

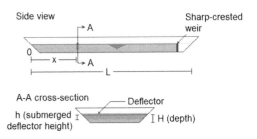

FIGURE 3.3   Side view of the unit shown in Figure 3.2 – in detail: deflector height (h/H) and position (x/L) relative to the water depth (H) and reactor length (L) (adapted from Almeida, 1997)

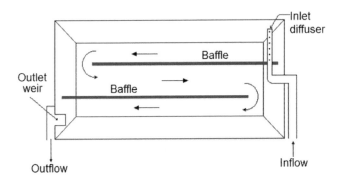

FIGURE 3.4   Top view of a WWTU with baffles as an example of complementary design elements aimed at promoting uniformity of flow characteristics (adapted from Almeida, 1997)

width (B) and increase the length (L) of the main flow path, from the inlet to the outlet section. Such dimensions are usually considered relative to one another in formulating the flow geometry indicator length-to-width ratio ($\beta = L/B$) (e.g. Thackston et al., 1987). Inserting baffles in a rectangular unit tends to create a serpentine-like flow path. A similar effect can be achieved with pipe reactors mounted in compacted way as shown by Wilson and Venayagamoorthy (2010).

For instance, Siqueira (1998) assessed the hydrodynamic performance of two setups of a WWTU physical model, as illustrated in Figure 3.5. The WWTU consisted of a rectangular unit in which baffles were introduced to enhance flow uniformity towards a plug flow-like flow pattern. Due to baffling, β increased from 2.3 in the reference setup to approximately 22 in the modified setup. The corresponding RTD curves are given in Figure 3.6. It can be noted that the use of baffles led to a flow pattern in the reactor closer to plug flow than observed in the unbaffled setup, in which the flow pattern tended to complete mixing.

Generally speaking, establishing a relatively high β value in some types of WWTU is usually aimed at favouring plug flow-like conditions. Where no particular attention has been paid to improve inlet and outlet design (e.g. Marske and Boyle, 1973), or in the case of very complex flow fields in WWTU (e.g. Teixeira, 1993), it is not uncommon to find recommended β values in the range of 20 to 40. However, other aspects also come into play at this point. While such an intervention tends to be a low risk conservative approach to hydrodynamic performance improvement (towards plug flow), it

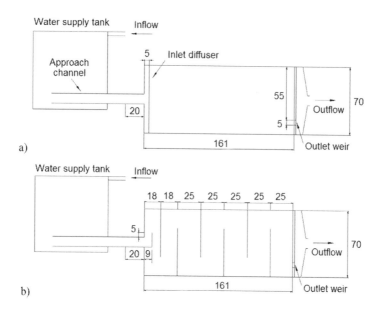

FIGURE 3.5   Effect of WWTU setup on the flow pattern, showing: a) reference unit with no baffles; and b) modified unit with baffles (dimensions in cm) (adapted from Siqueira, 1998)

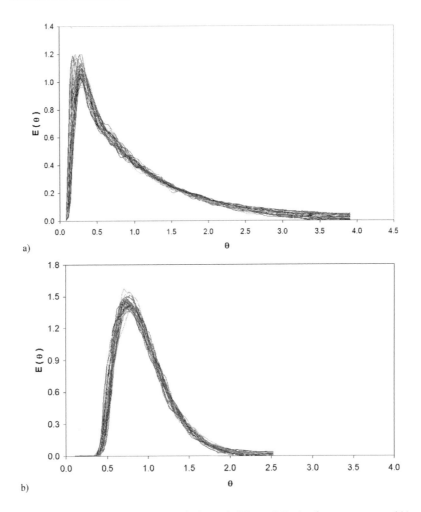

FIGURE 3.6 RTD curves for the unit shown in Figure 3.5: a) reference setup; and b) baffled setup (adapted from Siqueira, 1998; Teixeira and Siqueira, 2008)[5]

5 Republished with permission of American Society of Civil Engineers, from *Journal of Environmental Engineering*, Performance Assessment of Hydraulic Efficiency Indexes, Teixeira, Siqueira, 134 (10), 851–859, 2008; permission conveyed through Copyright Clearance Center, Inc.

may lead to unnecessarily high improvement costs, as discussed further in Chapter 4, and impair in other ways, for example access for maintenance. The international technical and scientific literatures abound in reports of the effect of baffles on WWTU hydrodynamic performance. In addition to early work cited in Chapters 1 and 2, interested readers can also refer to Almeida (1997) (also discussed below), Moreira (1999), Figueiredo (2000) (also discussed below), Stamou (2002) (referred to as guiding walls), Rauen (2005) (also discussed below), Abbas et al. (2006), Gualtieri (2007), Zhang et al. (2007), Goula et al. (2008), Stamou (2008), Gualtieri (2009), Kim et al. (2010), Wilson and Venayagamoorthy (2010), Wols et al. (2010), Amini et al. (2011), Lee et al. (2011), Olukanni and Ducoste (2011), Shahrokhi et al. (2012), Zhang et al. (2013), Angeloudis et al. (2014a), Taylor et al. (2015), Chang et al. (2016), Coggins et al. (2018), Demirel and Aral (2018) and Li et al. (2018).

3.3 INLET AND OUTLET DEVICES

Where other design and operational conditions can be optimised, such as those related to inlet and outlet characteristics, the need to promote high β values can be dramatically reduced – indeed, it is possible to achieve conditions relatively close to plug flow even with β values of the order of unity. For instance, Howe et al. (2012) recommended β values of at least 4 for sedimentation tanks, while von Sperling (1996b) suggested that β should be at least 2–4 for stabilisation ponds, with both types of units also being fitted with adequate inlet and outlet structures to promote flow uniformity around the cross-sectional area, from the inlet to outlet sections. As illustrated in Figures 3.1–3.5, a manifold, sharp-crested weir or dented plate is recommended as an outlet structure, while a diffuser can be fitted as an inlet device.

Similar recommendations for distributed inlet and outlet devices can also be found in design guidelines of flocculators (Stevenson, 1997; Libanio, 2010), stabilisation ponds (von Sperling, 1996b; Jordao and Pessoa, 2005) and anaerobic reactors (Chernicharo, 1997), among others. Further investigations of the effect of inlet design on WWTU hydrodynamic performance include Stamou and Noutsopoulos (1994), Rauen (2005) (also discussed below), Sozzi (2005), Greene et al. (2006), Rostami et al. (2011), Angeloudis et al. (2014b), Barnett and Venayagamoorthy (2014), Carlston and Venayagamoorthy (2015), Angeloudis et al. (2016) and Cruz et al. (2016).

3.4 FLOW DEFLECTORS

Introducing flow deflectors can help to distribute and dissipate momentum around the flow cross-section, particularly to counteract the effects of, for example, an inflowing jet, and where the aim is to achieve plug flow-like conditions. The primary reason for using deflectors in WWTU is, thus, to mitigate the formation of recirculation zones in the vertical and/or horizontal planes,[6] in such a way to contribute to reducing the overall intensities of short-circuiting and mixing in the WWTU.

An in-depth investigation of the effects of deflector design on WWTU flow patterns was reported by Almeida (1997) for a sedimentation unit,[7] with additional information being provided by Siqueira and Teixeira (1996), Teixeira et al. (1996) and Almeida et al. (1997). Varying deflector design features led to the identification of optimal height and positioning along the flow path in the unit. Such WWTU, which is illustrated in Figures 3.1–3.4, had a rectangular plan view with dimensions 87.6 m x 40.6 m and depth of 8.0 m. A 1:20 scaled down model of such a WWTU was assessed for a number of setups involving baffles and a flow deflector. Three discharge values were tested in the reference setup: Q_{min} = 1.0 L.s^{-1}, Q_{av} = 2.5 L.s^{-1} and Q_{max} = 3.7 L.s^{-1}. Figure 3.7 shows the passage curves for Q_{min}, Q_{av} and Q_{max}, as well as for the plug flow pattern taken as reference for analysis, i.e., considering the dispersion effects, as outlined in Section 2.3.2.

A visual direct comparison between the set of passage curves for the three flows rates[8] and the passage curve for the reference plug flow indicates that the original setup has low hydrodynamic efficiency: the t_i values approaching zero and the long tails of the passage curves point out that there exists in the unit high intensity of short circuit and presence of a large

---

6 Such as those exemplified in Figures 2.9, 2.10 and 2.11.

7 For high sedimentation efficiency, incoming flow parcels need to follow a trajectory long enough for settling to occur within the unit, ideally subjected to relatively low turbulence. Thus, short-circuit should be mitigated and the flow field should be as uniform as possible along the unit, which indicates that plug flow is the ideal flow pattern for this WWTU.

8 RTD curves obtained under different discharge values appear largely superimposed herein, rather than staggered along the horizontal axis. This is because both the tracer passage times (t) and the theoretical residence time (T), which define Θ, are affected by such flow rate change. This is a key effect of the normalisation promoted to achieve an RTD curve, i.e. to allow direct comparisons of curves obtained, potentially under different conditions, so that assessments can more easily focus on the relative spread of substance concentration. This aspect is discussed further in Section 3.6.

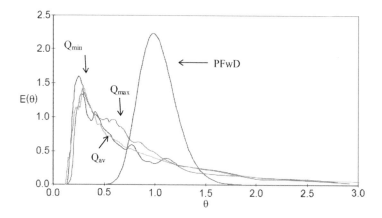

FIGURE 3.7   Tracer passage curves for PFwD, Q_{min}, Q_{av}, Q_{max} for the WWTU model illustrated in Figure 3.1 (adapted from Almeida, 1997)

portion of its volume occupied by dead zones and recirculating flows, thus contributing to the high degree of mixing observed.

The t_{10} values for the three conditions of flow rate are given in Table 3.1 and ranged from 0.24 to 0.28. When compared to the values of t_{10} corresponding to the ideal plug flow (0.82) and complete mixing (0.11), they inform that the flow in the unit tends to complete mixing. This trend can also be shown by means of the indicator σ^2 (see data in Table 3.1), whose values for the three flow rate conditions ranged from 0.64 to 0.82, differing greatly from the desired value of σ^2 (= 0.03 – for plug flow with dispersion) and approaching the value of σ^2 corresponding to complete mixing (= 1.0).

The results of both forms of assessment presented here (Qualitative – through comparative observation of passage curves; and Semi-quantitative – developed based on HEI for short-circuit intensity and degree of mixing) were consistent in

TABLE 3.1 Short-circuiting and mixing HEI values obtained by Almeida (1997) for the reference setup of a sedimentation unit and corresponding idealised flow results

HEI	PFwD	CM	Q_{min}	Q_{av}	Q_{max}
t_{10}	0.82	0.11	0.24	0.26	0.28
σ^2	0.03	1.00	0.72	0.82	0.64

indicating the low hydrodynamic efficiency associated with the original setup. As a first attempt to approximate the flow in the unit to the ideal plug flow with dispersion, as recommended by the literature (Hart, 1979; Falconer and Tebbutt, 1986), the flow deflector presented in Figures 3.2 and 3.3 was used. An investigation showed that the hydrodynamic performance of the unit was affected by the ratio between deflector height (h) and the flow depth (H), as well as by its location (x) in relation to the flow length (L).

For each flow rate (Q_{min}, Q_{av} and Q_{max}), the study considered three h/H values (0.2; 0.4; 0.6) and 5 or 6 positions for x/L (between 0.2250 and 0.4500), making a total of 50 test scenarios. Three indicators were also used in the comparative evaluation of the behaviour of the variation of the deflector design parameters h/H and x/L in relation to the hydrodynamic efficiency of the unit. They are:

1. Position of the set of passage curves corresponding to the unit setups with deflector, for a given flow value, relative to those passage curves to the original setup and for plug flow with dispersion. For each h/H value, Figure 3.8 shows, for Q_{min}, the resulting passage curves having the three highest t_{10} values. As can be observed, with the increase of the parameter h/H, the set of the three curves tends to approach the passage curve for plug flow with dispersion; or, in other words, the hydrodynamic efficiency of the unit also tends to increase.

2. Behaviour of t_{10} with the increase of x/L for each h/H value. Figure 3.9 presents the results for Q_{min}. Taking, initially, those referring to h/H equal to 0.4 and 0.6, it is observed that with the increase of x/L, t_{10} tends to grow to a maximum value, then decreases. For h/H equal to 0.2, the results show a tendency to increase the value of t_{10} with the increase of x/L, and to reach a maximum around x/L equal to 0.4500.

3. Position of t_{10} for each pair of values (x/L; h/H) in relation to the HEI values for the original setup and for plug flow with dispersion. In this case, Figure 3.9 shows that: a) for any setup scenario with deflector investigated, for the minimum flow rate, Q_{min}, the short-circuiting levels were lower than that referring to the original setup – that is, the values resulting from t_{10} were greater than 0.24; and b) the pair of values (x/L; h/H) that led to the highest hydrodynamic efficiency was (0.3375; 0.6).

After performing a similar analysis for the two other flow rates (Q_{av} and Q_{max}), the overall result was that:

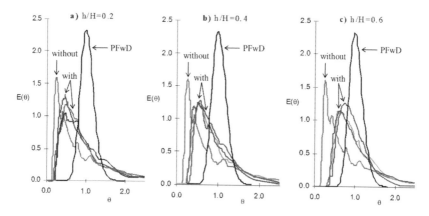

FIGURE 3.8 Tracer passage curves without and with flow deflector for Q_{min} with three relative deflector heights (h/H) and at various relative locations (x/L) (adapted from Almeida, 1997)

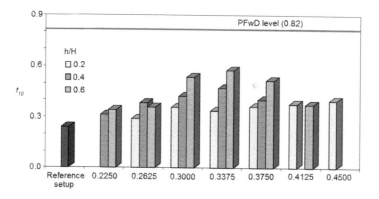

FIGURE 3.9 Results for t_{10} obtained with the variation of deflector position (x/L) for experiments conducted under Q_{min} for three deflector heights (h/H) (adapted from Almeida, 1997)

- For all three flow rates ($Qmin$, Qav and $Qmax$), the deflector height $h/H = 0.6$ was the most effective in reducing the short-circuiting levels associated with the original setup, without deflector.
- Optimum position for deflector installation is located in the range of $0.3000 – 0.3375$ for x/L.

- Despite the improvement in hydrodynamic efficiency promoted with the implementation of the flow deflector in the unit's original setup, the results demonstrate the limitation of this hydrodynamic performance correction device in WWTU, requiring the adoption of complementary corrective actions aiming at making closer the flow pattern in the unit to plug flow with dispersion.

Other investigations of the effect of flow deflectors on WWTU hydrodynamic performance include Falconer and Tebbutt (1986) (referred to as submerged weirs), Lyn and Rodi (1990), Cestari et al. (2012) and Demirel and Aral (2018) (referred to as horizontal baffles).

3.5 RELATIVE IMPACT OF FLOW MODIFIER DEVICES

As presented in Section 3.4, the flow pattern in the sedimentation unit assessed by Almeida (1997) had considerable intensities of short-circuiting and mixing, as depicted from Figure 3.7, and the use of a flow deflector was not, by itself, so effective to make the flow in the unit satisfactorily close to the ideal flow of reference – plug flow with dispersion. Thus, an investigation was conducted on the impact of the use of baffles in the improvement of the hydrodynamic efficiency of the original setup. The unit with baffles assessed is that shown in Figure 3.4, which employs two longitudinal baffles, i.e. parallel to the longest side of the tank, resulting in a significant increase in the β ratio value, from 2.5 to 23.8. The tracer passage curves obtained for the three investigated flow rates for this setup are illustrated in Figure 3.10.

In Figure 3.10 it can be noted that t_i for the passage curves referring to the setup with baffles closely approximate t_i for the passage curve corresponding to plug flow with dispersion; this being a strong indication of the baffles' effectiveness in reducing the short-circuiting levels associated with the original setup. This can be corroborated by the result of the comparative assessment of the t_{10} values in Table 3.2 for the baffled setup (0.66 to 0.75) compared to those for the original setup (0.24 to 0.28) and for plug flow with dispersion (0.82), which indicates that there is a quite high level of proximity between the flow pattern in the baffled unit and plug flow.

It can also be seen from Figure 3.10 that the passage curves for the original setup have high asymmetry with respect to the vertical axis passing by the peaked concentration and well-elongated tails, while the passage

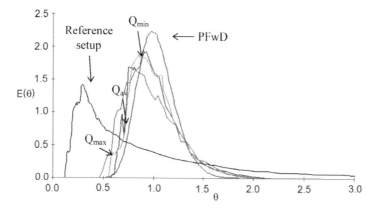

FIGURE 3.10 Tracer passage curves obtained for the baffled setup (adapted from Almeida, 1997)

curves referring to the baffled setup show relatively low level of asymmetry and short tails. These indicate that the installation of the two baffles in the original setup caused a large reduction in the mixing levels in the unit. This indication can be confirmed by the proximity between the values of σ^2 in Table 3.2 for the baffled setup (0.06 to 0.08) and for plug flow with dispersion (0.003).

The results of these evaluations indicate that the use of the two long-itudinal baffles proved to be more effective than the flow deflector in reducing the high short-circuiting levels and degrees of mixing associated with the original setup.

Teixeira et al. (1997), Figueiredo (2000) and Figueiredo and Teixeira (2000) investigated the effects of baffling and flow deflector for an unusual L-shaped model WWTU. The reference setup (unbaffled, no flow deflector)

TABLE 3.2 HEI results for the reference and baffled setups (adapted from Almeida, 1997)

HEI	PFwD	REFERENCE SETUP			BAFFLED SETUP		
		Q_{min}	Q_{av}	Q_{max}	Q_{min}	Q_{av}	Q_{max}
t_{10}	0.82	0.24	0.26	0.28	0.74	0.75	0.66
σ^2	0.03	0.72	0.82	0.64	0.07	0.06	0.08

of such a unit was scaled-down from a field scale contact unit used in a water treatment plant located in the Brazilian southeast region. As illustrated in Figure 3.11, four distinct setups were thus investigated in the laboratory using tracer tests, namely the reference setup, a setup with flow deflector (with variable height comparable to inlet tube diameter), a setup with baffles and a setup with flow baffles and a different flow deflector. The flow deflectors were installed right 'in front' of the pipe inlet section and were aimed at dissipating and quickly diffusing momentum of the jet-like inflow. An intended consequence of this was to mitigate the occurrence of extensive vertical recirculating flow in the unit, which otherwise tended to dominate the flow pattern and cause high short-circuiting and mixing levels. The baffling arrangement increased β from 3.7 to 33.

RTD curves for the investigated setups are shown in Figures 3.12 and 3.13, while Table 3.3 contains the HEI and d values associated with such an assessment. It can be initially noted, from Figure 3.12, that the flow pattern in

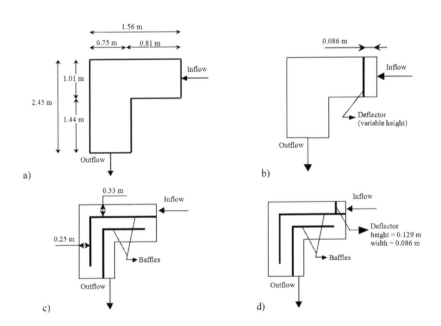

FIGURE 3.11 Plan views of L-shaped model contact unit setups assessed by Figueiredo (2000) using tracer techniques, showing: a) reference setup; b) setup with flow deflector; c) setup with baffles; and d) setup with baffles and flow deflector (adapted from Figueiredo, 2000; Figueiredo and Teixeira, 2000)

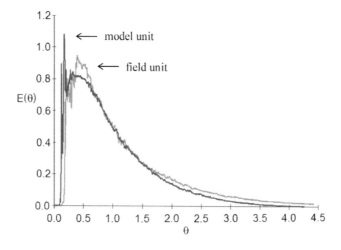

FIGURE 3.12   RTD curves measured in the field scale and model L-shaped WWTU under the reference setup (unbaffled, no flow deflector) (adapted from Teixeira et al., 1997; Figueiredo, 2000)

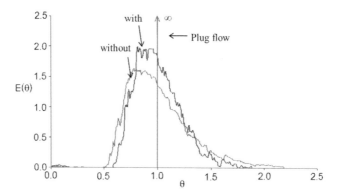

FIGURE 3.13   RTD curves measured in the model L-shaped WWTU under the setups with baffles and flow deflector and with baffles without flow deflector, as well as for plug flow (adapted from Figueiredo, 2000)

TABLE 3.3 Tracer test results (HEI and *d* values) obtained for the investigated setups and idealised flow patterns complete mixing and plug flow (adapted from Figueiredo, 2000)

HEI	COMPLETE MIXING	REFERENCE SETUP	WITH FLOW DEFLECTOR	WITH BAFFLES	WITH BAFFLES AND FLOW DEFLECTOR	PLUG FLOW
t_i	0.0	0.11	0.23	0.55	0.58	1.0
t_{10}	0.11	0.26	0.44	0.72	0.74	1.0
σ^2	1.0	0.51	0.31	0.12	0.12	0.0
Mo	21.8	7.5	4.2	2.3	2.1	1.0
d	∞	0.407	0.191	0.065	0.065	0.0

the reference setup tended to complete mixing, which was undesirable, since plug flow is the ideal flow pattern for such a WWTU type. It was this very factor that motivated the detailed assessment undertaken by Figueiredo (2000).

These results indicated that the setup with baffles and flow deflector conferred the best hydrodynamic performance to the WWTU assessed by Figueiredo (2000), among the options thus investigated. This is because the corresponding RTD curve and HEI values were the closest ones to the corresponding plug flow results (with plug flow being the ideal flow pattern for contact units). Judging from the HEI results, the improvements achieved by using a flow deflector were moderate and small for the reference and baffled setups respectively. The *d* values included in Table 3.3 were used in process efficiency simulation, as presented and discussed in Chapter 4.

In addition to the presence of baffles and other flow-modifier structures, their quantity, dimensions and arrangement inside a WWTU can have a substantial impact on the flow pattern. Take, for example, the model contact unit setups investigated by Rauen (2005), as illustrated in Figures 3.14 and 3.15, covering five baffling arrangements, two baffling orientations and two types of inlet devices. Baffle dimensions were such that allowed for a nearly-uniform flow width along the serpentine path, apart from the last compartment in setups MS1 and MS2. As shown in Figure 3.14b, the inlet devices comprised a near surface channel (C) and a near bed pipe (P) in compartment 1, which were used alternately in experimentation with setups OS and MS4. The corresponding tracer test results are shown in Figure 3.15 and Table 3.4.

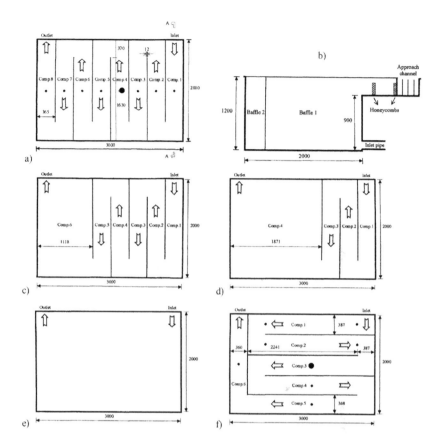

FIGURE 3.14 Schematic diagrams of model WWTU setups investigated by Rauen (2005), showing: a) horizontal plan view of the original setup (OS) with eight compartments; b) vertical plan view of inlet devices (A-A cross-section in pan a); c) horizontal plan view of modified setup MS1 with six compartments; d) horizontal plan view of modified setup MS2 with four compartments; e) horizontal plan view of modified setup MS3 with one compartment; and f) horizontal plan view of modified setup MS4 with six compartments and a mixed orientation of baffles (dimensions in mm) (dots indicate locations of velocity profiles shown in Figure 3.16) (adapted from Rauen, 2005)

As can be noted from Figure 3.15 and Table 3.4, the gradual introduction of baffles in the setup order MS3 → MS2 → MS1 → OS caused the flow pattern to change from close to complete mixing to close to PFwD, owing

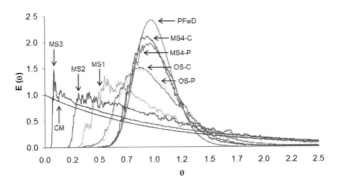

FIGURE 3.15 RTD curves obtained for the investigated setups and idealised flow patterns complete mixing (CM) and plug flow with dispersion (PFwD) (adapted from Rauen, 2005)

TABLE 3.4 Tracer test results (HEI and d values) obtained for the investigated setups and idealised flow patterns complete mixing (CM) and plug flow with dispersion (PFwD) (adapted from Rauen, 2005)

SETUP OR FLOW PATTERN CODE	t_i	t_{10}	Mo	σ^2	d
PFwD	0.50	0.81	1.52	0.027	0.014
MS4-P	0.50	0.79	1.71	0.052	0.028
MS4-C	0.48	0.78	1.69	0.055	0.029
OS-C	0.47	0.70	2.12	0.095	0.050
OS-P	0.30	0.68	2.13	0.097	0.053
MS1	0.28	0.50	3.38	0.224	0.132
MS2	0.19	0.37	4.89	0.306	0.190
MS3	0.05	0.16	11.5	0.539	0.451
CM	0.00	0.11	21.8	1.00	$\rightarrow \infty$

primarily to a concomitant reduction of reactor volume occupied by recirculating flow in the horizontal plane. The corresponding extreme β values were 1.5 (setup MS3) and 44 (setup OS). The type of inlet device also affected the results, but discrepancies were relatively small compared to the effect of increasing the number of compartments (as depicted from a comparison between results for setups OS-C and OS-P). Irrespective of the inlet type, setup MS4 provided the closest condition to the PFwD flow pattern, even though it had a similar β value (40) as setup OS. This was explained primarily by the $90°$ change in orientation of the first baffle in setup MS4, which then also exerted the effect of a deflector for the inflow jet.

Mean velocity measurements undertaken by acoustic Doppler velocimetry (ADV) in the model unit by Rauen (2005) are shown in Figure 3.16. Vertical velocity profiles measured at the centre of each compartment of setup OS (locations indicated by dots in Figure 3.14a) are illustrated in igures 3.16a (compartments one to four) and 3.16b (compartments five to eight), while vertical profiles measured in setup MS4 (locations indicated by dots in Figure 3.14f) are shown in Figures 3.16c (compartments one and two, with two profiles each) and 3.16d (compartments three to six, with one profile each). A comparison can be made of profiles measured at similar relative flow lengths, such as the profile measured in compartment four of setup OS (situated at a relative flow length of 0.437 – larger dot in Figure 3.14a) and in compartment three of setup MS4 (situated at a relative flow length of 0.434 – larger dot in Figure 3.14f). While the former was still substantially affected by three dimensional flow (as illustrated for a similar unit in Figures 2.11 and 2.12), the latter was relatively uniform thanks to the baffle that divided compartments one and two also acting as an inflow deflector device in the first compartment, spreading and attenuating kinetic energy of the incoming flow.

Other than that, the type of baffling orientation[9] may also have played a role. While transversal baffling was used in setup OS, setup MS4 comprised mostly longitudinal baffling (compartments one to five), with transversal baffling being verified only in compartment six. As shown by Teixeira (1993), the practical effect expected from such a design modification would be for setup MS4 to have less reactor volume occupied by recirculating flow in the horizontal plane in comparison with setup OS. Hence, it can be concluded that achieving flow uniformity in a shorter space inside setup MS4 was due to a combined effect of reducing recirculating flows in both the vertical and horizontal planes inside the unit.

---

9 Where transversal (or cross) baffling is defined, for a rectangular unit, as that in which compartments are oriented at a $90°$ angle in relation to the longest unit dimension, while longitudinal baffling is that in which compartments are in line with such dimension (Teixeira, 1993).

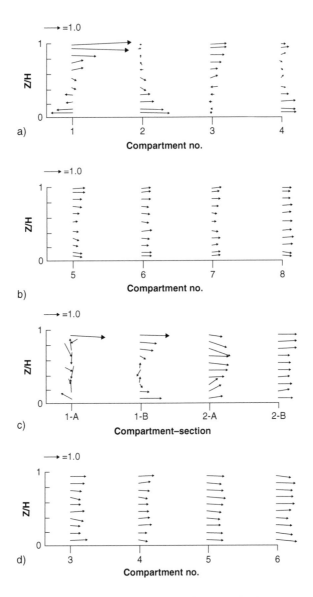

FIGURE 3.16 Vertical profiles of mean velocity fields obtained by Rauen (2005) for setup OS (Figures a and b) and setup MS4 (Figures c and d), where Z/H is the normalised height above the bed and the length of the reference velocity vector is shown at the top left corner of each plot (Rauen, 2005)

Investigations of the effects of other types of flow modifier devices can be found in Hannoun and Boulos (1997) (for perforated plates referred to as diffusion walls), Greene et al. (2006) (for perforated plates), Lee et al. (2011) (for perforated plates referred to as diffuser walls), Tarpagkou and Pantokratoras (2014) (for lamella plates), Angeloudis et al. (2016) (for a variety of devices including rounded corners, baffle tip modification, guiding vanes and an inclined bed), Soukane et al. (2016) (for rounded corners and spiral design) and Kizilaslan et al. (2018) (for perforated plates referred to as porous baffles).

3.6 A NOTE ON THE RELATIVE IMPACT OF FLOW RATE VARIATION ON WWTU HYDRODYNAMIC PERFORMANCE

As shown in Figures 3.7 and 3.10 and Tables 3.1 and 3.2, the dimensionless RTD curves and associated HEI results were not significantly impacted by the flow rate variation promoted by Almeida (1997). As a result, if any one of the flow rate values were to be taken as the design flow rate, the expected hydrodynamic performance would be comparable. It can be noted that changing the prototype unit setup had a stronger impact on the hydrodynamic efficiency than varying the flow rate, as concluded from the analysis of the Hydrodynamic Efficiency Indicators.[10]

Similar conclusions, pointing to a virtually negligible impact of discharge variation on WWTU hydrodynamic performance, were derived by Figueiredo (2000), Teixeira et al. (2000), Rauen (2001) and Machado (2002) (also discussed in Teixeira and Rauen, 2014), provided that the flow regime in the unit is turbulent and that the flow pattern is not close to plug flow. For non-turbulent flow and as the flow pattern approaches plug flow, then a discharge reduction tends to cause higher short-circuiting and mixing levels in the unit, as also shown by Mitha and Mohsen (1990), Teixeira and Sant'Ana (1999) and Barnett and Venayagamoorthy (2014). This occurs as the relative importance

---

10 By contrast, any discharge variation can potentially affect the mean contact time and, thus, process efficiency in a significant way even if normalised HEI values remain virtually unchanged, as shown by Rauen (2001) and Freitas et al. (2005). For example, increasing the discharge tends to enhance advective (streamwise) solute transport and, thus, reduce the mean residence time in a WWTU – which can potentially reduce process efficiency if other parameters are constant.

of the viscous resistance force increases *vis-à-vis* the other key flow driving forces (e.g. gravitational and/or pressure forces). Such a force balance change creates an effect akin to loss of dynamic similitude between such a lower discharge condition and its reference discharge counterpart, as shown by Teixeira and Rauen (2014) for scale effects in WWTU physical models. Although the range of discharge values assessed in such studies may be deemed relatively narrow, it tends to cover typical operating conditions of the units so that such conclusions do have practical implications.

Higher discharge values generally also induce higher levels of turbulence, which may not be beneficial for certain treatment processes, such as sedimentation. These aspects are illustrated based on mean velocity and turbulence intensity fields in Figures 3.17–3.20 for the WWTU model shown in Figures 2.7–2.9, as reported by Teixeira (1993). The unit (Tank.2 in Figure 2.8) was assessed under two discharge values, namely Q_{min} = 2.72 L.s^{-1} and Q_{max} = 6.54 L.s^{-1}, which gave mean cross-sectional flow velocities of U_0 = 0.6 cm.s^{-1} and U_0 = 1.5 cm.s^{-1} respectively. Point velocity results were normalised by such U_0 values in each condition, leading to the velocity vector plots shown in Figures 3.17 and 3.18 for compartments one and two of the unit respectively. Point turbulence intensities u' were also normalised by U_0 and used to generate the contour plots of Figures 3.19 and 3.20 for compartments one and two respectively of the unit.

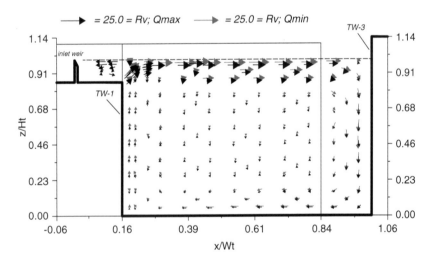

FIGURE 3.17 Mean normalised velocity fields measured in the mid-width longitudinal-vertical plane of compartment one under two discharge values (Q_{min} and Q_{max}) (Teixeira, 1993)

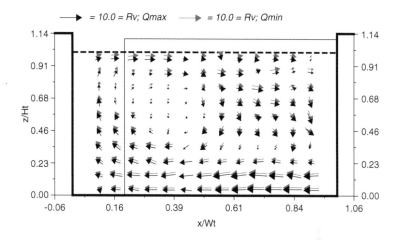

FIGURE 3.18 Mean normalised velocity fields measured in the mid-width longitudinal-vertical plane of compartment two under two discharge values (Q_{min} and Q_{max}) (Teixeira, 1993)

It can be noted in Figures 3.17 and 3.18 that the overall relative flow patterns in compartments one and two under distinct discharge values were similar, apart from localised discrepancies in magnitude near the water surface in compartment one and near the bed in compartment two – which were caused by slightly different inflow conditions in the approach channel. As a result, the relative levels of turbulence were also greater under Q_{min} in compartment one, particularly near the water surface (as shown in Figure 3.19) and in compartment two, particularly near the bed (as shown in Figure 3.20).

3.7 CONCLUDING REMARKS

This chapter illustrated how hydrodynamics of reactors can offer practical solutions that can lead to the improvement of treatment processes. Depending on the WWTU type, requirements other than those associated with the RTD of fluid particles appear as design recommendations aimed at improving process performance. Take, for instance:

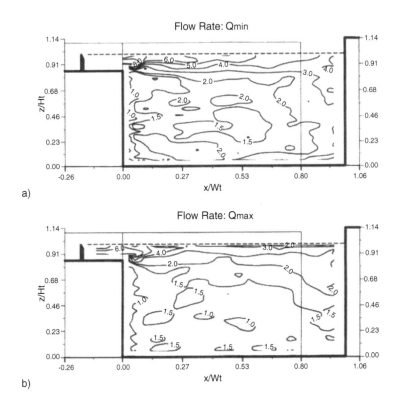

a)

b)

FIGURE 3.19 Normalised turbulence fields measured in the mid-width longitudinal-vertical plane of compartment one under two discharge values: a) Q_{min} and b) Q_{max} (Teixeira, 1993)

- Flocculation units: these WWTUs have internal flow requirements associated with inducing high enough shear and turbulence levels to promote particle collisions (so that flocs can form and grow) and prevent settling within the flocculator (Stevenson, 1997), but not superior to a given threshold, above which formed flocs could break (Oliveira, 2014; Oliveira and Teixeira, 2017; Vaneli and Teixeira, 2019). Outlet design recommendations are also made so that nappe flow shear does not exceed a certain threshold that could equally cause formed flocs to break up and, thus, decrease the overall flocculation efficiency of the unit. The design of flocculation units typically relies on a specific type of hydro-kinetic model, one which includes the so-called velocity gradient in the flow. However, the

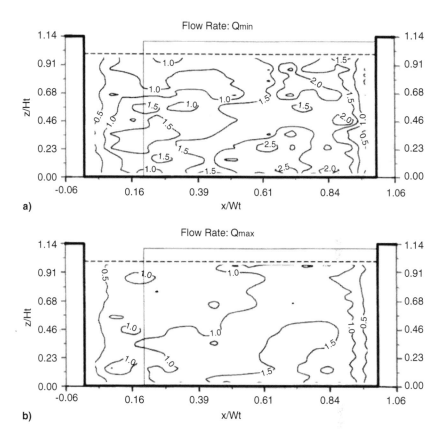

FIGURE 3.20 Normalised turbulence fields measured in the mid-width longitudinal-vertical plane of compartment two under two discharge values: a) Q_{min} and b) Q_{max} (Teixeira, 1993)

standard way to determine such a parameter involves only mean flow quantities and simplifying assumptions in terms of flow hydrodynamics (e.g. Camp and Stein, 1943; Richter, 2009). Furthermore, Stevenson (1997) argues that plug flow in hydraulic flocculators is not necessarily ideal, as that would prevent larger flocs from colliding against smaller ones, which in theory would be situated upstream along the flow path, and which can accelerate the process of the unit as a whole. The closest theoretical model for this is perhaps the tanks-in-series model including recirculation with throughflow,[11] i.e. a series of complete mixing flow chambers that gives, as a whole, the desired mean residence time where floc size would correlate with chamber number, with retro-feed of a portion of fluid from a later chamber (containing larger flocs) into an earlier chamber (containing smaller flocs). In this sense, Oliveira (2014), Sartori (2015) and Vaneli and Teixeira (2019) showed how the flocculation efficiency can be favoured by a combination of cross-sectional flow mixing and longitudinal dispersion within a single unit – so called helically coiled tube flocculator – which is currently at an experimental design stage.

- Sedimentation units: in these WWTUs, the occurrence of relatively tranquil flow with low turbulence levels favours the settling process, as turbulence favours vertical (as well as horizontal) diffusion of particles or flocs in the water column, which effectively delays their settling. Turbulent shear is also a key mechanism associated with causing the resuspension of settled material (sediment). It is thus undesirable to operate sedimentation units with flow velocities that cause relatively high turbulence levels – even if the flow is uniformly distributed in the cross-section and a mean HRT of interest is observed – as that could ultimately impair the overall unit process efficiency. Appropriate design guidelines should then be followed when available to avoid high turbulence levels in the sedimentation unit (e.g. Howe et al., 2012).

While relatively simple surrogate hydrodynamic performance indicators have been widely and successfully applied in WWTU design, as mentioned above, more complex and arguably better ones have been proposed in the international literature (e.g. Oliveira and Teixeira, 2019). Many such developments have only been made possible by the use of advanced CFD techniques, as well as more in-depth case-by-case assessments normally associated with design refinement and optimisation (e.g. Sartori et al., 2015; Angeloudis et al., 2016; Gao and Stenstrom, 2018). This is a growing and exciting field of interdisciplinary research, but one which is arguably still far from becoming standard practice in the water industry worldwide. Implementing and operating such advanced level CFD modelling requires highly skilled and trained professionals, which is still far from the reality of many, such as parts of the developing world where the struggle is still to universalise sewage collection and reach at least secondary treatment prior to disposal onto rivers or coastal water bodies. For this reason, important gains in terms of improving levels of water and wastewater quality

---

11 In Levenspiel (1999)'s terms.

more widely can still be achieved with more basic, but proven and cost-effective hydrodynamic design approaches.

The context for using assessment methods such as exemplified in this chapter and a structured approach for assessing inter-relations between hydrodynamic efficiency and process efficiency in WWTU are the subject of the next chapter.

Rational Procedures for Hydrodynamic Assessment and Design of WWTU

4

This chapter contains a guideline for assessing the impact of hydrodynamics on the treatment process performance in WWTU, both for existing and planned units. In each case, a reference setup and at least one alternative setup are considered, where the alternative setup is aimed at providing improved treatment process and/or hydrodynamic performance. The type of assessment covered herein is based on tracer passage curves, which can be obtained in field, laboratory and/or computational studies.

As mentioned in Chapter 1, to adequately assess WWTU performance one must take into account not only the treatment process efficiency, but also associated direct costs and the generation of undesirable by-products. In this context, at once broader and more complex, we strive here to offer guidance for enabling the application of knowledge on the hydrodynamics of reactors in the above mentioned assessment, which depends on:

- Which questions should be answered.
- Which treatment process performance indicators should be used.
- How the inter-relation between hydrodynamics and treatment process performance is undertaken as part of the assessment.

This section draws on practical case studies that encompass at least one performance indicator of the treatment process, such as the process efficiency. Two methods are used to establish the inter-relation of hydrodynamics and process performance, namely 1) direct method – when an analytical hydro-kinetic model is available to describe a given treatment process (i.e. one that explicitly includes in its formulation a hydrodynamic efficiency indicator that can be determined from tracer passage curves for the unit); and 2) indirect method – based solely on Hydrodynamic Efficiency Indicators for the WWTU.

4.1 APPLICATION OF DIRECT METHOD OF PERFORMANCE ASSESSMENT OF TREATMENT PROCESSES

The material in this section is based on practical scenarios selected to represent the following key treatment process efficiency (TPE) conditions: 1) WWTU efficiency for a reference design setup with real flow [(TPE)$_{RS-RF}$], determined through monitoring or estimated using a suitable modelling approach; 2) WWTU efficiency for a modified design setup with real flow [(TPE)$_{MS-RF}$], typically estimated using a suitable modelling approach; 3) WWTU efficiency for a modified design setup with ideal flow [(TPE)$_{RS-IF}$], typically estimated using a suitable modelling approach; and 4) minimum required or desired WWTU efficiency [(TPE)$_{Min}$] (e.g. arising from regulation or an internal demand for the treatment system).

4.1.1 Condition where (TPE)$_{RS-RF}$ ≥ (TPE)$_{Min}$

For such a condition, even if a required efficiency for the treatment process is met by the WWTU reference setup, decision makers may wish to know whether, and how much, improving the WWTU hydrodynamic efficiency (possibly achieved by altering the reference design setup) would provide significant improvement of the treatment process performance. If this is the case, the sequence of steps given below may provide technical guidance and a foundation for the decision-making process. A hypothetical example of the water disinfection process by chlorination in a contact tank and the hydro-kinetic model of Wehner and Wilhelm (1956)[1] (Equations 1.1 and 1.2) are used:

- Step 1 – Determination of treatment process efficiency gain [ΔTPE] in case the WWTU hydrodynamic efficiency of a modified design setup approaches the hydrodynamic efficiency related to the corresponding ideal flow condition.

---

1 Models such as this are used herein for illustration purposes only, without necessarily implying their validity and effectiveness. A discussion and comparison of different models used to simulate the water disinfection process by chlorine in contact tanks can be found in Teixeira et al. (2016).

$$\Delta TPE = (TPE)_{RS-IF} - (TPE)_{RS-RF} \qquad (4.1)$$

Equations 1.1 and 1.2 can be applied to this sample scenario, where $(TPE)_{RS-IF} = DE$, with C_{out}/C_{in} calculated using the values of k and T corresponding to $(TPE)_{RS-RF}$, while the value of d can be estimated from the plug flow with dispersion flow pattern. As shown in Section 2.3.2, such estimation requires determining the following parameters: Manning coefficient (n); effective flow length (L); wetted area (A) and perimeter (P) of the flow cross section; and flow depth (H). Alternatively, d can be obtained using physical or computational modelling of the planned WWTU setup modification aimed at achieving a plug flow-like flow pattern, such as by using baffles, with or without a flow deflector.

✓ Step 2 – Verification of the significance of ΔTPE on the overall treatment process performance.

A first aspect of interest here refers to which performance factors, besides ΔTPE, will be taken into account in this verification. In the case of water disinfection by chlorination, possible factors of interest include: increased maintenance and infrastructure costs (IMIC), reduced operation costs (ROC), reduced generation of by-products undesirable to the treatment process (RGBTP) and to human health (RGBHH). In this instance, it is not relevant to consider environmental impacts in the analysis.

The decision on which of these factors will be included in the function used to determine the overall treatment process performance (ΔTPP), and the nature of such function, will be taken in such a way as to provide technical support to decision makers, with possible approaches including optimisation and multi-criteria analysis methods. The latter has not usually figured in technical and scientific publications with the purpose suggested herein, but deserves attention for being closer to the decision-making reality in many countries, which are based on possibility ranges within acceptable negotiation margins for stakeholders, and not necessarily implies optimisation. For instance, it is possible to have a situation in which improving the hydrodynamic efficiency of a WWTU leads to significantly improved results for ΔTPE, ROC, RGBTP and RGBHH at the expense of increasing IMIC, but, even so, the required modification to the WWTU proves unfeasible due to insufficient funds for IMIC. This is a typical problem for multi-criteria analysis but not for optimisation methods.

The following real life example is used to illustrate how relatively small efforts in modifying WWTU reference design setups (increase in the value of IMIC) can lead to significant direct gain for the operational cost reduction

(ROC), while maintaining the same level for the treatment process efficiency. Comments on indirect gain of RGBHH will also be made.

The chlorine contact unit assessed by Teixeira, et al. (1997) and Figueiredo (2000) is considered for a comparison between the process performance of its reference and baffled setups, as illustrated in Figures 3.11a and 3.11c respectively. The disinfection efficiency (TPE) is related to the inactivation of total coliform, with key data as per Table 4.1.

As a starting point it is assumed that the reference setup operates with TPE = 99.99% as sufficient to meet the regulatory requirement of absence of total coliform in the outflow (as assessed in 100 mL samples collected at the outlet section of the unit), considering the characteristics of the source waters (Fonseca, 2002). In such a scenario, the intention is to quantify the impact of design modification on the operating cost of the WWTU as associated with disinfectant consumption.

As a result of making $DE = 99.99\%$ (= prescribed TPE), Equation 1.1 gives $C_{out}/C_{in} = 0.0001$. With $T = V/Q = 19.25$ min obtained from Equation 3.1 and Table 4.1, and $d = 0.407$ (Table 4.1), solving Equation 1.2 for C gives $C = C_{RS} = 0.390$ mg.L^{-1} for the reference setup, as the disinfectant concentration required for achieving the TPE value specified above.

The hydrodynamic performance improvement provided by baffling in such an instance led to $d = 0.065$ (Table 4.1), which then gives $C = C_{MS} = 0.123$ mg.L^{-1} to achieve the same TPE value using Equation 1.2 and other data, as per previous paragraph.

Since the expenditure with chlorine consumption in each contact unit setup is also a function of the volume of water that flows through the unit in any given period and the disinfectant cost – both of which are assumed, in

TABLE 4.1 Key characterising parameters of the disinfection process (adapted from Teixeira, et al., 1997; Figueiredo, 2000)

Reaction rate constant (k) (min^{-1})	$k = 4.02C^{0.801}$; where C is the disinfectant concentration (mg.L^{-1})
Reactive volume of each unit (m^3)	1.340
Average discharge (L.min^{-1})	69.6
Global mixing coefficient (d)	Reference setup: 0.407 Baffled setup: 0.065

this simulation, to be identical for each setup – then $[(C_{RS} - C_{MS})/C_{RS}] \times 100$ represents the relative reduction in ROC to achieve the same TPE value. Thus, ROC = 68.5% is obtained in this example, which indicates that a substantial cost reduction is reached due to the introduction of baffles in the WWTU.

It can also be highlighted that such ROC value would have a further positive effect of reducing the formation of chlorination by-products, such as potentially carcinogenic substances (RGBHH). This is particularly important in a region where the source water is rich in organic matter (Fonseca, 2002), which favours the formation of, for example, trihalomethanes during chlorination. Applying concepts of the area of hydrodynamics of reactors, such as the residence time distribution function, can lead to a more realistic estimate of the rate of trihalomethane formation in comparison with a more traditional approach, using the theoretical hydraulic residence time as the reaction time for all by-product formation, in models such as those found in Amy et al. (1987) and Brown et al. (2010).

4.1.2 *Condition where (TPE)$_{RS-RF}$ < (TPE)$_{Min}$*

In this case, one of the key questions to ask is can an improvement in hydrodynamic efficiency alone lead to meeting the minimum required efficiency? The answer to such a question can be based on a comparison of (TPE)$_{Min}$ and (TPE)$_{RS-IF}$. The following steps would be involved in such an analysis, considering the hypothetical situation introduced in Section 4.1.1:

✓ Step 1 – Determination of (TPE)$_{RS-IF}$ using a similar procedure as described in Section 4.1.1/Step 1.

✓ Step 2 – Comparison of (TPE)$_{Min}$ and (TPE)$_{RS-IF}$.

• If (TPE)$_{RS-IF}$ ≥ (TPE)$_{Min}$ then meeting the minimum required treatment process efficiency may occur solely through improving WWTU hydrodynamic efficiency, for example by modifying its geometry in such a way as to provide the required efficiency gain. If, furthermore, (TPE)$_{RS-IF}$ is significantly higher than (TPE)$_{Min}$, then two possible alternatives can be considered:

■ Maximise the treatment process efficiency by maximising hydrodynamic efficiency, i.e. achieving TPE = (TPE)$_{MS-IF}$. Such a condition will also maximise maintenance infrastructure costs and minimise operation costs and rates of by-product generation.

- Improve hydrodynamic efficiency only up to the point of achieving $(TPE)_{Min}$. Such a condition will minimise maintenance and infrastructure costs but not operation costs or the rates of by-product formation.

In practice, intermediate alternatives between such extremes can be considered as part of the decision-making process.

The following example illustrates how improving the hydrodynamic performance of a contact unit operating under a certain hydro-kinetic condition can lead to meeting the drinking water regulatory standard (in this case, for water disinfection) $(TPE)_{Min}$ of absence of total coliform in the outflow (as assessed in 100 mL samples collected at the outlet section of the unit). The WWTU setups and simulation condition considered herein are the same as mentioned in Section 4.1.1, including data provided in Table 4.1.

A starting point for this analysis is the assumption that $C = 0.123$ mg.L^{-1} is the maximum acceptable disinfectant (chlorine gas) dosage (for financial reasons) related to this treatment process stage. In such a scenario, the intention is to determine whether or not the reference setup provides sufficient disinfection efficiency to meet the corresponding regulatory standard $[(TPE)_{RS,RF} \geq (TPE)_{Min}]$; if it does not, then the goal is to determine whether such a requirement can be met using the baffled setup $[(TPE)_{MS,RF} >= (TPE)_{Min}]$. The average concentration of total coliform in the source water is taken as $C_{in} = 1000$ MPN.100 mL^{-1}, so that $(TPE)_{Min} = 99.90\%$ for this simulation.

The first question above can be tackled by firstly using Equations 1.1 and 1.2 with values of C_{in}, $T = 19.25$ min, $C = 0.123$ mg.L^{-1} and $d = 0.407$ for the reference setup to calculate TPE $= DE = 99.56\%$. As $(TPE)_{RS,RF} < (TPE)_{Min}$, it can be concluded that the reference setup operating under such a condition is unable to meet the drinking water standard considered herein.

The second question above can be tackled by comparing TPE $= 99.99\%$ obtained for the baffled setup under such condition (as obtained in Section 4.1.1) with $(TPE)_{Min}$. Since $(TPE)_{MS,RF} > (TPE)_{Min}$, the corresponding drinking water standard is met using the baffled contact unit setup, in this case.

If, on the other hand, $(TPE)_{MS-RF} < (TPE)_{Min}$ then meeting the minimum required treatment process efficiency cannot be achieved solely through improving hydrodynamic efficiency, i.e. additional measures such as increasing disinfectant dosage will be required. However, if maximising hydrodynamic efficiency provides a substantial gain in terms of the treatment process efficiency, then it may be worthwhile to modify the WWTU setup to achieve such an improvement. The procedure described in Section 4.1.1 can be used to assist this decision-making process.

This situation can be exemplified as follows, building on the case study outlined above. It is firstly assumed that the disinfectant dosage is $C = 0.035$ mg.

L^{-1}, which gives TPE = DE = 94.41% for the reference setup, using Equations 1.1 and 1.2 and all other data as above. In such a scenario of $(TPE)_{RS,RF}$ < $(TPE)_{Min}$, the intention is to determine whether implementing the baffled setup is sufficient to meet the regulatory standard, or whether additional measure(s) are required, such as increasing the disinfectant dosage.

Equations 1.1 and 1.2 can be used to calculate TPE = DE = 98.50% for such a condition, which gives $(TPE)_{MS-RF}$ < $(TPE)_{Min}$. Hence, meeting the regulatory standard in this instance requires additional measure(s) besides the hydrodynamic performance improvement provided by baffling. In this instance, another simulation with Equations 1.1 and 1.2 indicates that increasing the disinfectant dosage to C = 0.076 mg.L^{-1} in the baffled setup provides the required treatment efficiency level of $(TPE)_{Min}$ = 99.90%, i.e. the condition where $(TPE)_{MS-RF}$ = $(TPE)_{Min}$ is reached.

A critical assessment of the impact of hydrodynamic performance improvement on the disinfectant dosage requirement and associated cost reduction (ROC) – related to using the baffled setup *vis-à-vis* the reference setup – can be based on a comparison between the corresponding chlorine dosages required to meet the regulatory standard. Reaching TPE = 99.90% with the reference setup requires C = 0.205 mg.L^{-1}, a dosage which is 2.7 times higher than the corresponding value associated with the baffled setup.

4.2 APPLICATION OF INDIRECT METHOD OF PERFORMANCE ASSESSMENT OF TREATMENT PROCESSES

Applying the so-called indirect assessment method can be done in situations where the hydrodynamic performance and treatment process efficiency are positively and monotonically correlated,[2] i.e. where improving the former always enhances the latter. Examples that fall into this category include

---

2 The hydrodynamic performance and treatment process efficiency are not positively and mono-tonically correlated when the latter grows with the former only up to a point, which is then followed by a process efficiency reduction with any further improvement in hydrodynamic performance, as assessed using HEI. Using the indirect method discussed herein is unfeasible when such an inflection point cannot be determined, such as due to a lack of monitoring results for the process efficiency variation and/or a lack of suitable hydro-kinetic models.

water disinfection in contact tanks and pre-treatment substance homogenisation in equalisation units.

A procedure for using the indirect assessment method includes four steps, as outlined below.

✓ Step 1 – Determination of Hydrodynamic Efficiency Indicators (HEI) from the passage curve for the unit $[E(\theta)]_{RF}$ through field, laboratory or CFD experimentation, as outlined in Sections 2.2 and 2.3.2. The key HEI of interest are the short-circuit indicators $(t_i)_{RF}$ and $(t_{10})_{RF}$ and the mixing indicator $(Mo)_{RF}$.[3]

✓ Step 2 – Determination of HEI values from the passage curves for ideal flows, such that:

- For complete mixing, as shown in Table 2.1, $(t_i)_{IF/CM} = 0.0$; $(t_{10})_{IF/CM} = 0.11$; $(Mo)_{IF/CM} = 21.8$.
- For plug flow:

  ○ If the ideal flow pattern for the treatment process under analysis is complete mixing, then the $E(\theta)$ ideal (no dispersion) plug flow curve can be used in the analysis ($[E(\theta)]_{IF/PF}$), so that the indicator values are $(t_i)_{IF/PF} = (t_{10})_{IF/PF} = (Mo)_{IF/PF} = 1.0$.
  ○ If the ideal flow pattern for the treatment process under analysis is plug flow, then determine $E(\theta)$ for the plug flow with dispersion (PFwD) condition, e.g. using Equation 1.5 to obtain $[E(\theta)]_{IF/PFwD}$ with d determined as outlined in Section 4.1.1/Step 1. Indicator values $(t_i)_{IF/PFwD}$, $(t_{10})_{IF/PFwD}$ and $(Mo)_{IF/PFwD}$ can be determined from $[E(\theta)]_{IF/PFwD}$. If a $[E(\theta)]_{IF/PFwD}$ curve cannot be generated, then use the $[E(\theta)]_{IF/PF}$ curve and corresponding HEI values.

✓ Step 3 – Hydrodynamic efficiency assessment.

- For a qualitative assessment: plot together the $E(\theta)$ passage curves for the real and idealised flow patterns plug flow (with or without dispersion) and complete mixing (e.g. Figures 2.3 and 3.15).
- For a semi-quantitative assessment (more recommended due to its quantitative nature): tabulate the HEI values for t_i, t_{10} and Mo for the real and idealised flow patterns (e.g. Tables 3.1–3.4).

---

3 In this case, using σ^2 does not contribute significantly to the assessment undertaken using Mo and is inconvenienced by the fact that σ^2 is not recommended for low mixing conditions – see Section 2.3.2.

- Based on one or both such approaches, verify how close or distant the real flow features are from those of the idealised flow pattern for the treatment process under analysis (as per examples in Chapter 3).

  ◦ If distant (low hydrodynamic efficiency), then a proposition may be made to modify the unit setup in such a way as to improve its hydrodynamic efficiency;

    ▪ If plug flow is the idealised flow pattern of interest and it is possible to obtain the PFwD $E(\theta)$ curve, then the hydrodynamic efficiency assessment involving alternate setups should be more realistic than using the ideal plug flow scenario (e.g. Table 2.2).

    ▪ If complete mixing is the idealised flow pattern of interest, or if plug flow is the flow pattern of interest but it is not possible to obtain the PFwD $E(\theta)$ curve, then the analysis of alternate setups aimed at achieving the required hydrodynamic efficiency improvement does not benefit from having a realistic reference condition to indicate the theoretical limit of improvement, which tends to require a higher level of abstraction to the assessment. In such instances, theoretical and practical experience in similar assessments might be invaluable to direct refinements and further testing.

    ▪ If, on the other hand, the flow features are relatively close, then it is possible that no setup modification aimed at improving hydrodynamic efficiency is necessary (e.g. Figure 3.10 and Table 3.2).

✓ Step 4 – Hydrodynamic efficiency *versus* treatment process efficiency.

For a given setup taken as reference, it is important to assess whether improving hydrodynamic efficiency will significantly improve the treatment process performance. This is not an easy question to answer in a prognostic study by using the indirect method, due to the fact that no definitive answer can be obtained before corrective measures are implemented and tested. This is a consequence of the very nature of the method, which does not include means for estimating the treatment process efficiency, normally achieved by using mathematical hydro-kinetic models (analytical or others) and/or physical and computational modelling of the treatment process.

Nonetheless, the assessment team can advance by working in a multidisciplinary/transdisciplinary manner to enable consistent and rational decision making, as highlighted in the Preface. To exemplify how far the indirect method can go in comparison with the direct method, in terms of contributions to the treatment process performance, a hypothetical scenario of water disinfection by chlorination in a contact tank will again be used, with similar boundary conditions as described in Section 4.1.

4.2.1 Condition where $(TPE)_{RS-RF} \geq (TPE)_{Min}$

As mentioned in Section 4.1.1, this is a relatively comfortable condition where meeting the prescribed treatment level is not at risk – at least under current regulation or internal demand of the treatment process.

In this case, using the indirect method may not be enough to answer the key question of interest, i.e. whether an improvement in hydrodynamic efficiency will prove satisfactory in terms of treatment process performance, not only whether or not there will be an improvement, due primarily to its lack of a suitable tool to quantify the impact of hydrodynamic efficiency variation on TPE. Nonetheless, for such situations the following arguments can be considered:

a. In general, where the hydrodynamic efficiency is low for the reference design setup, one can expect to achieve significant TPE increase as WWTU modifications induce flow pattern improvement towards the idealised flow pattern for the process under analysis.

b. At the other extreme condition, as the hydrodynamic efficiency is relatively close to its peak condition (i.e. as HEI values approach those of the idealised flow pattern of interest), one can expect that further improvements in hydrodynamic efficiency do not significantly alter TPE.

c. The higher the departure from such two extreme conditions, the more difficult it is to ascertain the effect of hydrodynamic efficiency change on TPE using only the indirect method.

The example included in Section 4.1.1 can be used herein to assess the robustness of the strategy outlined in item "a" above. If used to assist decision-making on whether or not to implement measures aimed at improving the hydrodynamic efficiency of a WWTU, introducing baffles in the unit led to savings of 68.5% in terms of disinfectant consumption, as required to achieve a given TPE level. If a flow deflector is used instead (as per Figure 3.11b) to achieve $d = 0.191$ (as per Table 3.3), then the required disinfectant dosage $C = 0.223$ mg.L^{-1} is estimated using Equations 1.1 and 1.2. This leads to a 43% reduction in disinfectant consumption to achieve the same TPE level, thanks to a relatively straightforward design modification to the unit under consideration.

A related practical example can be used to illustrate item "b", above. As part of the planning stage of a hydrodynamic improvement study of such a contact unit, the aim is to quantify the TPE improvement and disinfectant consumption associated with promoting a flow pattern that gets closer and closer to plug flow, having DE = 99.99% as the target TPE. This is undertaken by hypothetically increasing the number of baffles in the unit, while also altering the inlet section to improve inflow cross-sectional distribution for each setup. The following relationship between β and σ^2, developed by Teixeira (1993) for transversal baffling in serpentine contact units, is also considered valid for such a unit:

$$\sigma^2 = 0.4\beta^{-0.52} \tag{4.2}$$

The β values thus obtained were 33.7, 58.4 and 110 for the setups with two, three and four baffles respectively, which give σ^2 values of 0.064, 0.048 and 0.035 respectively using Equation 4.2. The corresponding d values obtained using Equation 2.13 are 0.033, 0.025 and 0.018 respectively. Table 4.2 summarises the results associated with this simulation and includes the corresponding values for the unbaffled setup and two-baffle setup (with non-optimal inflow). Disinfectant dosages required to achieve TPE = 99.99% were determined for all setups using Equations 1.1 and 1.2, and other data, as per previous examples. Such results, obtained with a hydro-kinetic model are used in this example of the indirect method *in lieu* of measured process efficiency results, which were unavailable for such a hypothetical example.

Based on the results of Table 4.2, it is possible to infer on the effect of the inlet condition on the global mixing levels in the unit with two baffles: the setup with non-optimal inflow had a higher level of mixing (d = 0.065) than its idealised inflow counterpart (d = 0.033), and this occurred despite the similarity of the corresponding β values (= 33.74). Such an inflow improvement for the two-baffle setup led to a reduction of 0.032 in d, which was associated with a reduction of 0.026 mg.L^{-1} or 21.1% in C (from 0.123 mg.L^{-1} to 0.097 mg.L^{-1}).

These results also indicate that increasing the number of baffles from two to four led to a reduction of 0.015 in d, which varied from 0.033 to 0.018 respectively. This was associated with a reduction of 0.012 mg.L^{-1} 12.4% in C, which varied from 0.097 mg.L^{-1} to 0.085 mg.L^{-1} respectively. By contrast, using two baffles *vis-à-vis* the unbaffled setup led to a reduction of 0.342 in d, which varied from 0.407 to 0.065. This was associated with a reduction of 0.267 mg.L^{-1} or 68.5% in C, which varied from 0.390 mg.L^{-1} to 0.123 mg.L^{-1} respectively. Thus, increasing the number of baffles from two to four, in this instance, had a lower relative impact than optimising the

TABLE 4.2 Parameter values used in the hypothetical example of the indirect method as applied where the hydrodynamic efficiency is relatively close to its peak condition

SETUP	β	σ^2 (EQ. 4.2)	d (EQ. 2.13)	C (mg.L^{-1})
Unbaffled – non-optimal inflow	3.56	—	0.407	0.390
Two baffles – non-optimal inflow	33.74	—	0.065	0.123
Two baffles – idealised inflow	33.74	0.0642	0.033	0.097
Three baffles – idealised inflow	58.42	0.0482	0.025	0.090
Four baffles – idealised inflow	110.11	0.0347	0.018	0.085

inflow condition for the two baffle setup, and both impacts were considerably lower than the effect of introducing two baffles in the originally unbaffled setup.

Applying the indirect method in this example, where d and C were indeed positively and monotonically correlated, indicated that ever improving the flow pattern towards plug flow does not necessarily have a strong enough practical impact on the treatment process parameters, as assessed herein in terms of the disinfectant consumption required to achieve similar TPE levels in the unit. Hence, from the perspective of an integrated cost–benefit analysis, it would be beneficial to reach a compromise between expenditure on design modifications, savings originating from them and other investment demands.

4.2.2 Condition where $(TPE)_{RS\text{-}RF} < (TPE)_{Min}$

In contrast to the previously outlined condition, this is not a comfortable scenario as TPE must be improved – as a result of hydrodynamic efficiency or reaction kinetics enhancements, or both – if the minimum treatment level is to be achieved. The situation is made more complex by the unavailability of a tool, such as a hydro-kinetic model, that can be used to estimate the effect of varying hydrodynamic efficiency on TPE, and, particularly, if even the idealised flow condition in the WWTU under analysis would allow for meeting the required minimum treatment process efficiency, i.e. whether or not $(TPE)_{RS\text{-}IF} \geq (TPE)_{Min}$.

What can be done in this condition, if one is to make the most of their hydrodynamics of reactors knowledge, is to try and achieve a flow condition as close as possible to the idealised flow pattern of interest for the WWTU under analysis, ideally relying on adequate designer experience and judgement and bearing in mind possible practical constraints, such as highlighted in Chapter 3. Then, as mentioned above for Step 4, only after the WWTU is operational in such new condition can $(TPE)_{RS-IF}$ be determined to inform decision making on which alternative to follow:

- If $(TPE)_{MS-RF} \approx (TPE)_{RS-IF} \geq (TPE)_{Min}$, then meeting the minimum treatment process efficiency required improvement only of the hydrodynamic efficiency. As outlined in Section 4.1.2, decision makers can opt to maintain such a maximum hydrodynamic performance or to modify the WWTU such that $(TPE)_{Min}$ is met while other constraints and requirements are also taken into account and provided for, based on, for example, multi-criteria analysis as mentioned in Section 4.1.1.
- If $(TPE)_{RS-IF} < (TPE)_{Min}$, then meeting the minimum treatment process efficiency may be achieved on improving both the hydrodynamic efficiency and treatment process kinetics (e.g. by increasing reagent dosage), as mentioned in Section 4.1.2.

4.3 CONCLUDING REMARKS

The indirect method's effectiveness can only be confirmed after any modifications to the WWTU under analysis are implemented and tested in terms of their effects on the treatment process efficiency.

The direct method is more capable of informing the decision-making process in diagnostic and prognostic assessments, and thus in guiding design enhancements aimed at improving hydrodynamic efficiency. The indirect method is limited in this context, even where the hydrodynamic efficiency and TPE are positively and monotonically correlated.

References

Abbas, H.; Nasra, R.; Seif, H. (2006). Study of waste stabilization pond geometry for the wastewater treatment efficiency. *Ecological Engineering*, 28, 25–34.

Adams, E. W.; Rodi, W. (1990). Modelling flow and mixing in sedimentation tanks. *ASCE Journal of Hydraulic Engineering*, 116 (7), 895–913.

Almeida, M. M. P. (1997). A study of the effects of flow deflectors and length-to-width ratio of flow on the hydraulic efficiency of a sedimentation basin. Dissertation (M.Sc. Environmental Engineering), Universidade Federal do Espírito Santo. (in Portuguese)

Almeida, M. M. P.; Siqueira, R. N.; Teixeira, E. C. (1997). Influence of flow deflector height and position and of the flow length-to-width ratio on the hydrodynamic behaviour of a sedimentation tank. Proceedings of 19th Congress of the Brazilian Association of Sanitary and Environmental Engineering. (in Portuguese)

Amini, R.; Taghipour, R.; Mirgolbabaei, H. (2011). Numerical assessment of hydrodynamic characteristics in chlorine contact tank. *International Journal for Numerical Methods in Fluids*, 67, 885–898.

Amy, G. L.; Chadik, P. A.; Chowdhary, Z. K. (1987). Developing models for predicting trihalomethane formation potential and kinetic. *Journal of the American Water Works Association*, 79 (7), 89–97.

ANA. (2017). Agência Nacional de Águas. Sewage Atlas – removing pollution in watersheds. Brasília: Distrito Federal. (in Portuguese)

Angeloudis, A.; Stoesser, T.; Falconer, R. A. (2014a). Predicting the disinfection efficiency range in chlorine contact tanks through a CFD-based approach. *Water Research*, 60, 118–129.

Angeloudis, A.; Stoesser, T.; Kim, D.; Falconer, R. A. (2014b). Modelling of flow, transport and disinfection kinetics in contact tanks. *Proceedings of the ICE Water Management*, 167 (9), 532–546.

Angeloudis, A.; Stoesser, T.; Gualtieri, C.; Falconer, R. A. (2016). Contact tank design impact on process performance. *Environmental Modelling and Assessment*, 21, 563–576.

Barnett, T. C.; Venayagamoorthy, S. K. (2014). Laminar and turbulent regime changes in drinking water contact tanks. *Journal of the American Water Works Association*, 106 (12), 561–568.

Barter, P. J. (2003). Investigation of pond velocities using dye and small drogues: a case study of the Nelson City waste stabilisation pond. *Water Science & Technology*, 48 (2), 145–151.

Bellamy, W. D.; Finch, G. R.; Haas, C. N. (1998). *Integrated disinfection design framework*. Denver, CO: AWWA Research Foundation and American Water Works Association. 66 p.

Brasil. (1997). Federal Law 9433/1997. National Water Resources Policy. (in Portuguese)

Brown, D.; West, J. R.; Courtis, B. J.; Bridgeman, J. (2010). Modelling THMs in water treatment and distribution systems. *Proceedings of the ICE Water Management*, 163 (WM4), 165–174.

Camp, T. R.; Stein, P. C. (1943). Velocity gradients and internal work in fluid motion. *Journal of Boston Society of Civil Engineering*, 30, 219–237.

Capodaglio, A. G.; Hlavínek, P.; Raboni, M. (2016). Advances in wastewater nitrogen removal by biological processes: state of the art review. *Ambiente & Agua*, 11 (2), 250–267.

Carlston, J. S.; Venayagamoorthy, S. K. (2015). Impact of modified inlets on residence times in baffled tanks. *Journal of the American Water Works Association*, 107 (6), 292–300.

Cestari, J. C.; Matsumoto, T.; Sobrinho, M. D.; Libanio, M. (2012). Hydrodynamic evaluation of a mechanical flocculation pilot unit. *Engenharia Sanitaria e Ambiental*, 17 (1), 95–106. (in Portuguese)

Chang, T. J.; Chang, Y. S.; Lee, W. T.; Shih, S. S. (2016). Flow uniformity and hydraulic efficiency improvement of deep-water constructed wetlands. *Ecological Engineering*, 92, 28–36.

Chernicharo, C. A. L. (1997). *Anaerobic reactors*. Belo Horizonte: Editora UFMG. 380 p. (in Portuguese)

Coggins, L. X.; Sounness, J.; Zheng, L.; Ghisalberti, M.; Ghadouani, A. (2018). Impact of hydrodynamic reconfiguration with baffles on treatment performance in waste stabilisation ponds: a full-scale experiment. *Water*, 10 (109), 1–18.

Crites, R. W.; Middlebrooks, E. J.; Reed, S. C. (2005). *Natural wastewater treatment systems*. Boca Raton: CRC Press/Taylor & Francis. 576 p.

Cruz, D. B.; Arantes, E. J.; Carvalho, K. Q.; Passig, F. H.; Kreutz, C.; Gonçalves, M. S. (2016). Hydrodynamic performance evaluation of an upflow anaerobic sludge blanket reactor with different configurations of the influent distribution system using computational fluid dynamics. *Engenharia Sanitaria Ambiental*, 21 (4), 721–730.

Danckwerts, P. V. (1953). Continuous flow systems – distribution of residence times. *Chemical Engineering Science*, 2 (1), 1–13.

Demirel, E.; Aral, M. M. (2018). Performance of efficiency indexes for contact tanks. *ASCE Journal of Environmental Engineering*, 144 (9), 04018076, 1–13.

Elder, J. W. (1959). The dispersion of marked fluid in turbulent shear flow. *Journal of Fluid Mechanics*, 5 (4), 544–560.

EPA. (1983). United States Environmental Protection Agency. Design manual – wastewater stabilization ponds. EPA-625/1-83-015, 342 p.

EPA. (1999). United States Environmental Protection Agency. Disinfection profiling and benchmarking guidance manual. EPA 815-R-99-013, 194 p.

Falconer, R. A.; Tebbutt, T. H. Y. (1986). A theoretical and hydraulic model study of a chlorine contact tank. *Proceedings of the ICE*, Part 2, 81, 255–276.

Figueiredo, I. C. (2000). A study on the influence of hydrodynamics and microorganism types on the chlorine dosage applied to water disinfection. Dissertation (M.Sc. Environmental Engineering), Universidade Federal do Espirito Santo. (in Portuguese)

Figueiredo, I. C.; Teixeira, E. C. (2000). Water disinfection – an evaluation of influences of unit setup and microorganism type on disinfectant consumption. Proceedings of 9th Luso-Brazilian Symposium on Sanitary and Environmental Engineering – SILUBESA. ABES / APRH, 2000, Porto Seguro - BA. (in Portuguese)

Fischer, H. B.; List, J. E.; Imberger, J.; Koh, C. R.; Brooks, N. H. (1979). *Mixing in inland and coastal waters*. San Diego: Academic Press. 483 p.

Fonseca, I. R. (2002). Assessing the impact of hydraulic performance on the disinfection efficiency of water chlorination contact tanks for total and faecal coliform groups. Dissertation (M.Sc. Environmental Engineering), Universidade Federal do Espirito Santo. (in Portuguese)

Freitas, K. R.; Fonseca, I. R.; Siqueira, R. N.; Teixeira, E. C. (2005). Influence of the flow rate on the disinfection process performance in continuous flow units. Proceedings of the 23rd Congress of the Brazilian Association of Sanitary and Environmental Engineering. (in Portuguese)

French, R. H. (1985). *Open-channel hydraulics*. Singapore: McGraw-Hill. 739 p.

Gao, H.; Stenstrom, M. K. (2018). Evaluation of three turbulence models in predicting the steady state hydrodynamics of a secondary sedimentation tank. *Water Research*, 143, 445–456.

Goula, A. M.; Kostoglou, M.; Karapantsios, T. D.; Zouboulis, A. I. (2008). A CFD methodology for the design of sedimentation tanks in potable water treatment: case study: the influence of a feed flow control baffle. *Chemical Engineering Journal*, 140, 110–121.

Greene, D. J.; Haas, C. N.; Farouk, B. (2006). Computational fluid dynamics analysis of the effects of reactor configuration on disinfection efficiency. *Water Environment Research*, 78 (9), 909–919.

Gualtieri, C. (2007). Analysis of the effect of baffles number on a contact tank efficiency with Multiphysics 3.3. Proceedings of the COMSOL Users Conference, 2007, Grenoble.

Gualtieri, C. (2009). Analysis of flow and concentration fields in a baffled circular storage tank. Proceedings of 33rd IAHR Congress, 2009, Vancouver.

Gyurek, L. L.; Finch, G. R. (1998). Modeling water treatment chemical disinfection kinetics. *Journal of Environmental Engineering, ASCE*, 124 (9), 783–793.

Hannoun, I. A.; Boulos, P. F. (1997). Optimizing distribution storage water quality: a hydrodynamic approach. *Applied Mathematical Modelling*, 21, 495–502.

Hart, F. L. (1979). Modifications for the chlorine contact chamber. *Journal of the New England Water Pollution Control Association*, 13 (2), 135–151.

Hart, F. L.; Allen, R.; Dialesio, J.; Dzialo, J. (1975). Modifications improve chlorine contact chamber performance. *Water and Sewage Works*, 122 (4), 73–75.

Hart, F. L.; Gupta, S. K. (1978). Hydraulic analysis of model treatment units. *ASCE Journal of the Environmental Engineering Division*, 104 (EE4), 785–798.

Honey, R. E.; Hershberger, R.; Donnelly, R. J.; Bolster, D. (2014). Oscillating-grid experiments in water and superfluid helium. *Physical Review E*, 89, 053016, 1–11.

Howe, K. J.; Hand, D. W.; Crittenden, J. C.; Trussell, R. R.; Tchobanoglous, G. (2012). *Principles of water treatment*. Sao Paulo: Cengage. 602 p. (in Portuguese).

Jordao, E. P.; Pessoa, C. A. (2005). *Domestic wastewater treatment*. 4th ed. Rio de Janeiro: ABES. 932 p. (in Portuguese)

Karpinska, A. M.; Bridgeman, J. (2016). CFD-aided modelling of activated sludge systems – a critical review. *Water Research*, 88, 861–879.

Keller, E.; Pires, E. C. (1998). *Stabilisation ponds – project and operation*. Rio de Janeiro: ABES. 244 p. (in Portuguese)

Khan, L. A.; Wicklein, E. A.; Teixeira, E. C. (2006). Validation of a three-dimensional computational fluid dynamics model of a contact tank. *ASCE Journal of Hydraulic Engineering*, 132 (7), 741–746.

Kim, D.; Kim, D. I.; Kim, J. H.; Stoesser, T. (2010). Large Eddy Simulation of flow and tracer transport in multichamber ozone contactors. *ASCE Journal of Environmental Engineering*, 136, 22–31.

Kizilaslan, M. A.; Demirel, E.; Aral, M. M. (2018). Effect of porous baffles on the energy performance of contact tanks in water treatment. *Water*, 10 (1084), 1–15.

Kothandaraman, V.; Southerland, H. L.; Evans, R. L. (1973). Performance characteristics of chlorine contact tanks. *Journal of the Water Pollution Control Federation*, 45 (4), 611–619.

Laurent, J.; Samstag, R. W.; Ducoste, J. M.; Griborio, A.; Nopens, I.; Batstone, D. J.; Wicks, J. D.; Saunders, S.; Potier, O. (2014). A protocol for the use of computational fluid dynamics as a supportive tool for wastewater treatment plant modelling. *Water Science and Technology*, 70 (10), 1575–1584.

Lee, S.; Shin, E.; Kim, S. H.; Park, H. (2011). Dead zone analysis for estimating hydraulic efficiency in rectangular disinfection chlorine contactors. *Environmental Engineering Science*, 28 (1), 25–33.

Levenspiel, O. (1999). *Chemical reaction engineering*. 3rd ed. New York: John Wiley & Sons. 704 p.

Li, M.; Zhang, H.; Lemckert, C.; Roiko, A.; Stratton, H. (2018). On the hydrodynamics and treatment efficiency of waste stabilisation ponds: from a literature review to a strategic evaluation framework. *Journal of Cleaner Production*, 183, 495–514.

Libanio, M. (2010). *Fundamentals of water quality and treatment*. 3rd ed. Campinas: Editora Atomo. 494 p. (in Portuguese)

Louie, D. S.; Fohrman, M. S. (1968). Hydraulic model studies of chlorine mixing and contact chambers. *Journal of the Water Pollution Control Federation*, 40 (2–I), 174–184.

Lyn, D. A.; Rodi, W. (1990). Turbulence measurements in model settling tank. *ASCE Journal of Environmental Engineering*, 116 (1), 3–21.

Machado, C. M. (2002). Hydrodynamic performance assessment of a flocculation/ sedimentation unit used in the removal of cyanides and its association with the treatment process efficiency. Dissertation (M.Sc. Environmental Engineering), Federal University of Espirito Santo. 100 p. (in Portuguese)

Marske, D. M.; Boyle, J. D. (1973). Chlorine contact chamber design – a field evaluation. *Water and Sewage Works*, 120 (1), 70–77.

Meister, M.; Winkler, D.; Rezavand, M.; Rauch, W. (2017). Integrating hydrodynamics and biokinetics in wastewater treatment modelling by using smoothed particle hydrodynamics. *Computers & Chemical Engineering*, 99, 1–12.

Mitha, S. A.; Mohsen, M. F. N. (1990). Scale effect on dispersion in chlorine contact chambers. *Canadian Journal of Civil Engineering*, 17 (2), 156–165.

Moreira, C. R. F. (1999). A study of the hydrodynamic behaviour of baffled units aiming for optimal water disinfection in continuous flow contact tanks. Dissertation (M.Sc. Environmental Engineering), Federal University of Espirito Santo. 132 p. (in Portuguese)

Oliveira, D. S. (2014). Proposition of a model for performance estimation of helically coiled tube flocculators. Thesis (D.Sc. Environmental Engineering), Universidade Federal do Espírito Santo. (in Portuguese)

Oliveira, D. S.; Teixeira, E. C. (2017). Experimental evaluation of helically coiled tube flocculators for turbidity removal in drinking water treatment units. *Water SA*, 43, 378–386.

Oliveira, D. S.; Teixeira, E. C. (2019). Swirl number in helically coiled tube flocculators: theoretical, experimental, and CFD modeling analysis. International Journal of Environmental Science and Technology, 16 (7), 3735–3744.

Olukanni, D. O.; Ducoste, J. J. (2011). Optimization of waste stabilization pond design for developing nations using computational fluid dynamics. *Ecological Engineering*, 37, 1878–1888.

Pereira, C. B.; Teixeira, E. C. (2002). Influence of methods of determination of the longitudinal dispersion coefficient (D_L) in the simulation of water quality in rivers due to instantaneous releases of potentially polluting constituents. 5th International Conference on Hydroinformatics, Cardiff, UK.

Pritchard, P. J. (2011). *Fox and McDonald's introduction to fluid mechanics*. 8th ed. Hoboken: John Wiley & Sons. 877 p.

Raboni, M.; Torretta, V.; Viotti, P.; Urbini, G. (2014). Pilot experimentation with complete mixing anoxic reactors to improve sewage denitrification in treatment plants in small communities. Sustainability, 6, 112–122.

Rauen, W. B. (2001). Effects of scale and discharge on the dynamic similitude and chlorine disinfection efficiency of scaled-down models of contact tanks. Dissertation (M.Sc. Environmental Engineering), Federal University of Espirito Santo. 100 p. (in Portuguese)

Rauen, W. B. (2005). Physical and numerical modelling of 3-D flow and mixing processes in contact tanks. Ph.D. Thesis, Cardiff University, UK.

Rauen, W. B.; Angeloudis, A.; Falconer, R. A. (2012). Appraisal of chlorine contact tank modelling practices. *Water Research*, 46, 5834–5847.

Rauen, W. B.; Lin, B.; Falconer, R. A.; Teixeira, E. C. (2008). CFD and experimental model studies for water disinfection tanks with low Reynolds number flows. *Chemical Engineering Journal*, 137, 550–560.

Richter, C. A. (2009). *Water treatment methods and technology*. Sao Paulo: Blucher. 340 p. (in Portuguese)

Rigo, D.; Teixeira, E. C. (1995). Using tracers in design and optimisation of water and wastewater treatment units. Proceedings 18° CBESA, Salvador (in Portuguese).

Rostami, F.; Shahrokhi, M.; Said, M. A. M.; Syafalni, R. A. (2011). Numerical modeling on inlet aperture effects on flow pattern in primary settling tanks. *Applied Mathematical Modelling*, 35, 3012–3020.

Sarikaya, H. Z.; Saatçi, A. M. (1987). Bacterial die-off in waste stabilization ponds. *ASCE Journal of Environmental Engineering*, 113 (2), 366–382.

Sartori, M. (2015). Analysis of viscous deformation and phase segregation in helically coiled tube flocculators and associations with torsion coefficient. Thesis (D.Sc. Environmental Engineering), Universidade Federal do Espírito Santo. (in Portuguese)

Sartori, M.; Oliveira, D. S.; Teixeira, E. C.; Rauen, W. B.; Reis, N. C. (2015). CFD modelling of helically coiled tube flocculators for velocity gradient assessment. *Journal of the Brazilian Society of Mechanical Sciences and Engineering*, 37, 187–198.

Sawyer, C. M.; King, P. H. (1969). The hydraulic performance of chlorine contact tanks. Proceedings of the 24th Industrial Waste Conference, Purdue University, West Lafayette, USA.

Shahrokhi, M.; Rostami, F.; Said, M. A. M.; Sabbagh, S. R.; Syafalni, Y. (2012). The effect of number of baffles on the improvement efficiency of primary sedimentation tanks. *Applied Mathematical Modelling*, 36, 3725–3735.

Shilton, A.; Bailey, D. (2006). Drouge tracking by image processing for the study of laboratory scale pond hydraulics. *Flow Measurement and Instrumentation*, 17, 69–74.

Shilton, A.; Kreegher, S.; Grigg, N. (2008). Comparison of computation fluid dynamics simulation against tracer data from a scale model and full-sized waste stabilization pond. *ASCE Journal of Environmental Engineering*, 134, 845–850.

Shiono, K.; Teixeira, E. C. (2000). Turbulent characteristics in a baffled contact tank. *Journal of Hydraulic Research*, 38 (6), 403–416.

Silva, S. A.; Mara, D. D. (1979). *Biological treatment of wastewaters: stabilisation ponds*. Rio de Janeiro: ABES. (in Portuguese)

Siqueira, R. N. (1998). Development and improvement of criteria for assessing the hydraulic efficiency and calculation of mixing coefficient in water and wastewater treatment units. Dissertation (M.Sc. Environmental Engineering), Universidade Federal do Espírito Santo. (in Portuguese)

Siqueira, R. N.; Reisen, V.; Teixeira, E. C. (1999). Evaluation of several hydraulic efficiency indices as a tool for WWTU performance assessment. Proceedings of 20th Congress of the Brazilian Association of Sanitary and Environmental Engineering. ABES, Rio de Janeiro – RJ. (in Portuguese)

Siqueira, R. N.; Teixeira, E. C. (1996). Utilization of deflectors to minimize the flow kinetic energy at the entrance of contact tanks. Proceedings of 3rd CREEM, Rio de Janeiro. (in Portuguese)

Soukane, S.; Ait-Djoudi, F.; Naceur, W. M.; Ghaffour, N. (2016). Spiral-shaped reactor for water disinfection. Desalination and Water Treatment, 57 (48–49), 23443–23458.

Sozzi, A. (2005). CFD and PIV investigation of UV reactor hydrodynamics. Dissertation (M.Sc. Chemical and Biological Engineering), University of British Columbia.

Stamou, A. I. (2002). Verification and application of a mathematical model for the assessment of the effect of guiding walls on the hydraulic efficiency of chlorination tanks. Journal of Hydroinformatics, 4 (4), 245–254.

Stamou, A. I. (2008). Improving the hydraulic efficiency of water process tanks using CFD models. Chemical Engineering and Processing, 47, 1179–1189.

Stamou, A. I.; Adams, E. W. (1988). Study of the hydraulic behaviour of a model settling tank using flow through curve and flow patterns. Report SFB 210/E/36, University Karlsruhe, Karlsruhe, Germany.

Stamou, A. I.; Noutsopoulos, G. (1994). Evaluating the effect of inlet arrangement in settling tanks using the hydraulic efficiency diagram. Water S.A., 20 (1), 77–84.

Stamou, A. I.; Rodi, W. (1984). Review of experimental studies of sedimentation tanks. Report SFB 210/E/2, University Karlsruhe, Karlsruhe, Germany.

Stevenson, D. G. (1997). Water treatment unit processes. London: Imperial College Press. 474 p.

Sykes, R. M.; Walker, H. W.; Weavers, L. K. (2003). Chemical water and wastewater treatment processes. In: Chen, W. F.; Liew, J. Y. R. (eds.) The civil engineering handbook 10-1–10-56. 2nd ed. Boca Raton: CRC Press.

Tarpagkou, R.; Pantokratoras, A. (2014). The influence of lamellar settler in sedimentation tanks for potable water treatment – a computational fluid dynamic study. Powder Technology, 268, 139–149.

Taylor, Z. H.; Carlston, J. S.; Venayagamoorthy, S. K. (2015). Hydraulic design of baffles in disinfection contact tanks. Journal of Hydraulic Research, 53 (3), 400–407.

Tchobanoglous, G.; Burton, F. L.; Stensel, H. D. (2003). Wastewater engineering: treatment and reuse. Metcalf & Eddy, Inc. 4th ed.. New York: McGraw-Hill. 1819 p.

Teixeira, E. C. (1993). Hydrodynamic processes and hydraulic efficiency of chlorine contact units. Ph.D. Thesis, University of Bradford, UK.

Teixeira, E.C. (1995a). Importance of reactor hydrodynamics in optimising drinking water disinfection processes – A critical analysis. Proceedings of 18th Congress of the Brazilian Association of Sanitary and Environmental Engineering. TII-026. (in Portuguese)

Teixeira, E. C. (1995b). Improvement of the hydraulic performance of a chlorine contact unit by means of cross-baffling. International Symposium on Technology Transfer – IAWQ/ABES, Salvador.

Teixeira, E. C.; Almeida, M. M. P.; Siqueira, R. N.; Rigo, D. (1996). Diagnostics and improvement of the hydraulic efficiency of a sedimentation basin by means of tracer studies. Proceedings of AIDIS Congress, 1996, Mexico City. (in Portuguese)

Teixeira, E. C.; Andrade, M. W. M.; Rauen, W. B.; Machado, C. M. (2000). Influence of the hydrodynamic behaviour of a secondary clarifier on the chemical removal of cyanides from coke plant effluent. Proceedings of XXVII Interamerican Congress of Environmental and Sanitary Engineering, Porto Alegre. (in Portuguese)

Teixeira, E. C.; Chacaltana, J. T. A.; Pacheco, C. G.; Siqueira, R. N. (2004). Use of flow-through curves to calibrate and validate numerical models of solute transport in contact tanks. Proceedings of the 6th International Conference on Hydroinformatics, Suntec City, Singapore.

Teixeira, E. C.; Figueiredo Breda, I. C.; Resende, M. B.; Carvalho Neto, E. S. (1997). Hydraulic efficiency diagnosis of a contact tank and assessment of corrective measures for reducing short-circuiting. Proceedings of the 12th Symposium of the Brazilian Association of Water Resources. Vitoria/ES, Brazil. (in Portuguese)

Teixeira, E. C.; Rauen, W. B. (2014). Effects of scale and discharge variation on similitude and solute transport in water treatment tanks. ASCE Journal of Environmental Engineering, 140, 30–39.

Teixeira, E. C.; Rauen, W. B.; Fonseca, I.; Figueiredo, I. (2016). Experimental testing of water disinfection models under varying hydraulic and kinetic conditions. Water Environment Research, 88 (6), 521–530.

Teixeira, E. C.; Sant'Ana, T. D. (1999). Using hydrodynamic models in aerated submerged biofilters. Proceedings of the 20th Congress of the Brazilian Association of Sanitary and Environmental Engineering. (in Portuguese)

Teixeira, E. C.; Shiono, K. (1994). Three-dimensional characterisation of the hydraulic structure of flow, transport and mixing of solutes in a water chlorination model unit. Proceedings of VI SILUBESA, Florianopolis, 232–254. (in Portuguese)

Teixeira, E. C.; Siqueira, R. C. N.; Rauen, W. B. (2002). Automation of acquisition and processing of flow-through curves. Proceedings of the 5th International Conference on Hydroinformatics, Cardiff, UK.

Teixeira, E. C.; Siqueira, R. N. (2008). Performance assessment of hydraulic efficiency indexes. ASCE Journal of Environmental Engineering, 134 (10), 851–859.

Thackston, E. L.; Shields, F. D.; Schroeder, P. R. (1987). Residence time distributions of shallow basins. *ASCE Journal of Environmental Engineering*, 113 (6), 1319–1332.

Thirumurthi, D. (1969). A break-through in tracer studies in sedimentation tanks. *Journal of the Water Pollution Control Federation*, 41 (11), R405–R418.

Thirumurthi, D. (1974). Design criteria for waste stabilisation ponds. *Journal of the Water Pollution Control Federation*, 46 (9), 2094–2106.

Trussel, R. R.; Chao, J. L. (1977). Rational design of chlorine contact facilities. *Journal of the Water Pollution Control Federation*, 49 (4), 659–667.

United States Environmental Protection Agency (USEPA). (1999). Disinfection profiling and benchmarking guidance manual. 194 p.

Vaneli, B. P. (2014). Valuing the use of hydrodynamics of reactors in the oil & gas and water treatment sectors: a comparative study. Final year project (Environmental Engineering), Universidade Federal do Espirito Santo. (in Portuguese)

Vaneli, B. P.; Teixeira, E. C. (2019). Refinement of a water turbidity removal efficiency model by helically coiled tube flocculators. Engenharia Sanitaria e Ambiental, 24 (4), 773–783. DOI: 10.1590/S1413-41522019180405. (in Portuguese)

Von Sperling, M. (1996a). *Fundamentals of sewage treatment*. Belo Horizonte: Editora UFMG. 211 p. (in Portuguese)

Von Sperling, M. (1996b). *Stabilisation ponds*. Belo Horizonte: Editora UFMG. 196 p. (in Portuguese)

Watters, G. Z. (1972). The hydraulics of waste stabilization ponds. Reports – Utah State University. Paper 18.

Wehner, J. F.; Wilhelm, R. H. (1956). Boundary conditions of flow reactor. *Chemical Engineering Science*, 6, 89–93.

White, F. M. (2011). *Fluid mechanics*. 7th ed. New York: McGraw-Hill. 863 p.

Wilson, J. M.; Venayagamoorthy, S. K. (2010). Evaluation of hydraulic efficiency of disinfection systems based on residence time distribution curves. *Environmental Science & Technology*, 44, 9377–9382.

Wols, B. A.; Hofman, J. A. M. H.; Uijttewaal, W. S. J.; Rietveld, L. C.; van Dijk, J. C. (2010). Evaluation of different disinfection calculation methods using CFD. *Environmental Modelling & Software*, 25, 573–582.

Zhang, G.; Lin, B.; Falconer, R. A. (2000). Modelling disinfection by-products in contact tanks. *Journal of Hydroinformatics*, 2 (2), 123–132.

Zhang, J.; Huck, P. M.; Anderson, W. B.; Stubley, G. D. (2007). A computational fluid dynamics based integrated disinfection design approach for improvement of full-scale ozone contactor performance. *Ozone Science and Engineering*, 29, 451–460.

Zhang, J.; Tejada-Martínez, A. E.; Zhang, Q. (2013). Hydraulic efficiency in RANS of the flow in multichambered contactors. *ASCE Journal of Hydraulic Engineering*, 139, 1150–1157.

Index